DraftSight 2019

For Beginners

Tutorial Books

For Technical Support, contact us at:

online.books999@gmail.com

Table of Contents

Introduction

Draftsight is a CAD software developed and marketed by Dassault Systemes. It can be used to create, edit, and inspect 2D drawings and design. It has many useful features that are available in expensive 2D CAD packages.

Scope of this Book

The ***Draftsight 2019 for Beginners*** book provides a learn-by-doing approach for users to learn Draftsight. It is written for students and engineers who are interested to learn Draftsight 2019 for creating drawing of components or anyone who communicates through technical drawings as part of their work. The topics covered in this book are as follows:

- **Chapter 1, "Introduction to Draftsight 2019"**, gives an introduction to Draftsight. The user interface and terminology are discussed in this chapter.

- **Chapter 2, "Drawing Basics,"** explores the basic drawing tools in Draftsight. You create simple drawings using the drawing tools.

- **Chapter 3, "Drawing Aids,"** explores the drawing settings that assist you in creating drawings.

- **Chapter 4, "Editing Tools,"** covers the tools required to modify drawing objects or create new objects using the existing ones.

- **Chapter 5, "Multi View Drawings,"** teaches you to create multi-view drawings standard projection techniques.

- **Chapter 6, "Dimensions and Annotations,"** teaches you to apply dimensions and annotations to a drawing.

- **Chapter 7, "Section Views,"** teaches you to create section views of a component. A section view is the inside view of a component when it is sliced.

- **Chapter 8, "Blocks, Attributes, and References,"** teaches you to create Blocks, Attributes, and references. Blocks are a group of objects in a drawing that can be reused. Attributes are notes or values related to an object. References are drawing files attached to another drawing.

- **Chapter 9, "Sheets and Annotative Objects,"** teaches you to create sheets and annotative objects. Sheets are the digital counterparts of physical drawing sheets. Annotative objects are dimensions, notes, and so on, which their sizes concerning drawing scale.

Chapter 1: Introduction to DraftSight 2019

In this chapter, you learn about:

- **DraftSight user interface**
- **Customizing user interface**

Introduction

Draftsight is a Computer-aided designing tool with many features. It can be used by architects, engineers, and construction service providers, as well as professional CAD users, designers, educators, and hobbyists. This book helps you to learn the basics of Draftsight.

System requirements

The following are system requirements for running DraftSight smoothly on your system.

Minimum

- **64-bit version**: Microsoft® Windows® 7 Service Pack 1, Windows 8.1, or Windows 10
- **32-bit version**: Microsoft Windows 7 Service Pack 1
- Intel® Core™ 2 Duo, or AMD® Athlon™ x2 Dual-Core processor
- 500 MB free hard disk space depending on accessory applications installed
- 2 GB RAM
- 3D Graphics accelerator card with OpenGL version 1.4
- Display with 1280 x 768 pixels resolution
- Mouse

Recommended

- **64-bit version**: Microsoft Windows 7 Service Pack 1, Windows 8.1, or Windows 10
- **32-bit version**: Microsoft Windows 7 Service Pack 1
- Intel Core i5 processor, AMD Athlon/Phenom™ x4 processor, or better
- 1 GB free hard disk space depending on accessory applications installed
- 8 GB RAM
- 3D Graphics accelerator card with OpenGL version 3.2 or better
- Full HD monitor
- Wheel mouse, or 3D mouse

Starting DraftSight 2019

To start **DraftSight 2019**, double-click the **DraftSight 2019** icon on your Desktop (or) click **Start > All apps > D > Dassault Systemes > DraftSight 2019**.

DraftSight user interface

When you double-click the DraftSight 2019 icon on the desktop, the DraftSight 2019 initial screen appears.

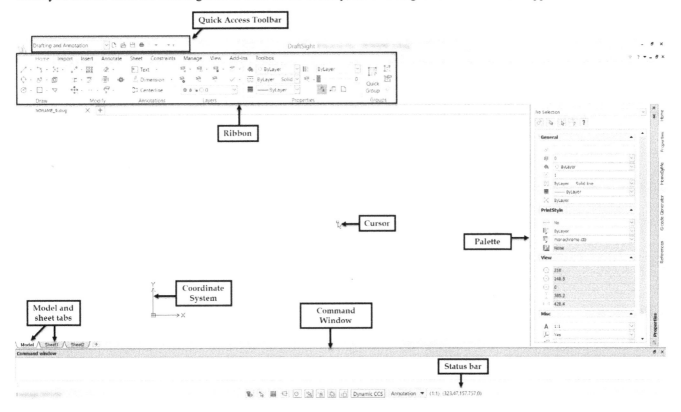

Workspaces in DraftSight

There are three workspaces available in DraftSight: **Drafting and Annotation** and **Classic**. By default, the **Drafting and Annotation** workspace is activated. You can create 2D drawings in this workspace. You can also enable other workspaces by using the **Workspace** drop-down on the top-left corner of the window.

*Tip: If the **Workspace** drop-down is not displayed at the top left corner, then click the down arrow next to Quick Access Toolbar. Next, select **Workspace** from the drop-down; the **Workspace** drop-down is visible on the Quick Access Toolbar.*

Drafting and Annotation Workspace

This workspace has all the tools to create a 2D drawing. It has a ribbon located at the top of the screen. The ribbon is arranged in a hierarchy of tabs, panels, and tools. Panels such as **Draw**, **Modify**, and **Layers** consist of tools that are grouped based on their usage. Panels, in turn, are grouped into various tabs. For example, the panels, such as **Draw**, **Modify**, and **Layers**, are located in the **Home** tab.

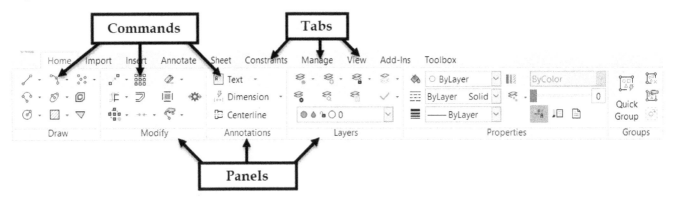

File Menu

The **File Menu** appears when you click on the icon located at the top left corner of the window. The **File Menu** consists of a list of open menus. You can see a list of recently opened documents on the right side.

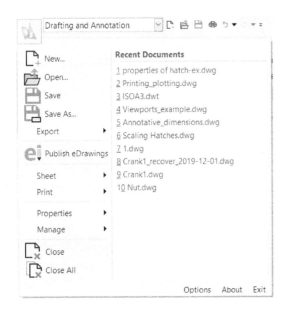

Quick Access Toolbar

The Quick Access Toolbar is located at the top left corner of the window and helps you to access commands quickly. It consists of commonly used commands such as **New**, **Save**, **Open**, **Print**, and **Redo**.

File tabs

File tabs are located below the ribbon. You can switch between different drawing files by using the file tabs. Also, you can open a new file by using the + button, easily.

Graphics area

The graphics area is the blank space located below the file tabs. You can draw objects in the graphics area.

Command Window

The Command Window is located below the graphics window. It is easy to execute a command using the Command Window. You can type the first letter of command, and it lists all the commands starting with that letter. The command window helps you to activate commands very quickly and increases your productivity.

Also, the Command Window shows the current state of the drawing. It shows various prompts while working with any command. These prompts are a series of steps needed to execute a command successfully. For example, when you activate the LINE command, the Command Window displays a prompt, "Specify start point." You need to click in the graphics window to specify the start point of the line. After specifying the start point, the prompt, "Specify next point" appears. Now, you need to specify the next point of the line. It is recommended that you should always have a look at the Command Window to know the next step while executing a command.

Status Bar

Status Bar is located at the bottom of the DraftSight window. It contains many buttons which help you to create a drawing very easily. You can turn ON or OFF these buttons just by clicking on them. The buttons available on the status bar are briefly discussed in the following section.

Button	Description
Annotation ▼ (1:1) (242.482,91.338,0) **Drawing Coordinates**	It displays the drawing coordinates when you move the pointer in the graphics window. You can turn OFF this button by right click on it and selecting OFF.
Snap (F9)	The Snap button aligns pointer only with the Grid points. When you turn ON this button, the pointer can select only the Grid points.
Grid (F7)	It turns the Grid display ON or OFF. You can set the spacing between the grid lines by right-clicking on the Grid button and selecting the **Settings** option. You can use grid points along with the Snap mode to draw objects easily and accurately.
Ortho (F8)	It turns the Ortho Mode ON or OFF. When the Ortho Mode is ON, only horizontal or vertical lines can be drawn.
Polar (F10)	This icon turns ON or OFF the Polar Tracking. When the Polar Tracking is turned ON, you can draw lines easily at regular angular increments, such as 5, 10, 15, 23, 30, 45, or 90 degrees. Notice a trace line displayed when the pointer is at a particular angular increment. You can set the angular increment by right-clicking on this button and selecting the **Settings** option. 43.148 < 45 °
ESnap (F3)	This icon turns ON or OFF the ESnap mode. When this mode is turned ON, you can easily select the key points of objects such as endpoints, midpoint, and center point, and so on.

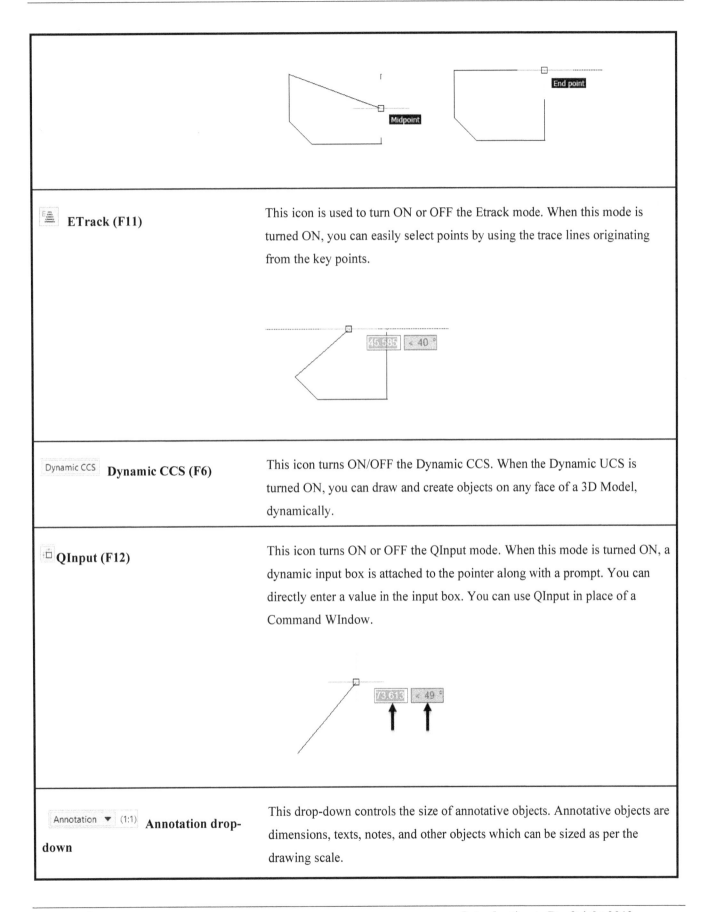

ETrack (F11)	This icon is used to turn ON or OFF the Etrack mode. When this mode is turned ON, you can easily select points by using the trace lines originating from the key points.
Dynamic CCS (F6)	This icon turns ON/OFF the Dynamic CCS. When the Dynamic UCS is turned ON, you can draw and create objects on any face of a 3D Model, dynamically.
QInput (F12)	This icon turns ON or OFF the QInput mode. When this mode is turned ON, a dynamic input box is attached to the pointer along with a prompt. You can directly enter a value in the input box. You can use QInput in place of a Command WIndow.
Annotation drop-down	This drop-down controls the size of annotative objects. Annotative objects are dimensions, texts, notes, and other objects which can be sized as per the drawing scale.

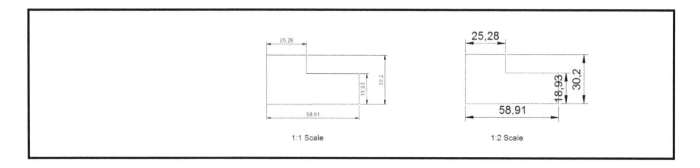

Minimizing/Maximizing the Ribbon

You can minimize the ribbon by clicking the arrow button located at the top right corner, as shown below.

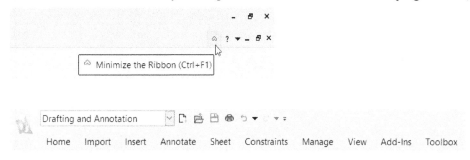

You can maximize the ribbon by again clicking the arrow button.

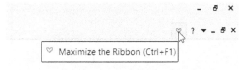

Dialogs and Palettes

Dialogs and Palettes are part of the DraftSight user interface. Using a dialog or a palette, you can easily specify many settings and options at a time. Examples of dialogs and palettes are as shown below.

Dialog

Palette

Shortcut Menus

Shortcut Menus appear when you right-click in the graphics window. DraftSight provides various shortcut menus to help you access tools and options very easily and quickly. There are various types of shortcut menus available in DraftSight. Some of them are discussed next.

Right-click Menu

This shortcut menu appears whenever you right-click in the graphics window without activating any command or selecting an object.

Select and Right-click menu

This shortcut menu appears when you select an object from the graphics window and right-click. It consists of editing and selection options.

Command Mode shortcut menu

This shortcut menu appears when you activate command and right-click. It shows options depending upon the active command. The shortcut menu below shows the options related to the RECTANGLE command.

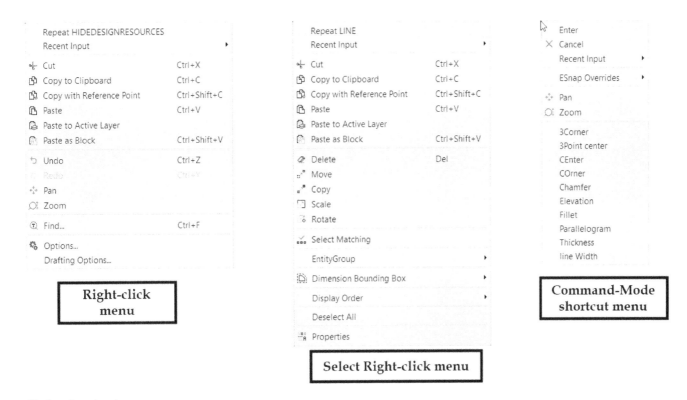

Grip shortcut menu

This shortcut menu is displayed when you select a grip of an object, move the pointer and right-click. It displays various operations that can be performed using grip.

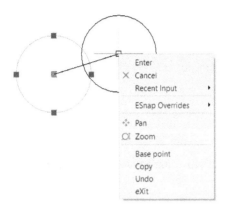

Selection Window

A selection window is used to select multiple elements of a drawing. You can select multiple elements by using two types of selection windows. The first type is a rectangular selection window. You can create this type of selection window by defining its two diagonal corners. When you define the first corner of the selection window on the left and second corner on the right side, the elements which completely fall under the selection window are selected.

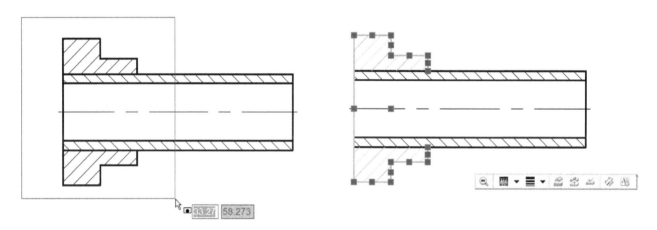

However, if you define the first corner on the right side and the second corner of the left side, the elements, which fall completely or partially under the selection window, are selected.

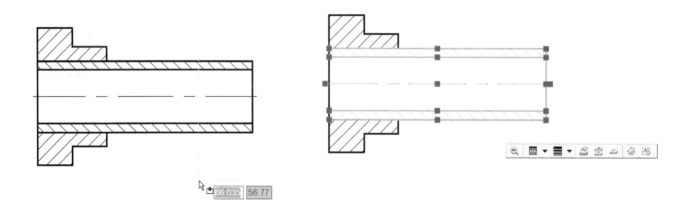

Starting a new drawing

You can start a DraftSight document by using the **Specify template** dialog.

To start a new drawing, click the **New** button on any one of the following:

- Quick Access Toolbar
- **File Menu**

The **Specify Template** dialog appears when you click the **New** button. In this dialog, select the **standard.dwt** (inch units) or **standardiso.dwt** (metric units) template for creating a drawing.

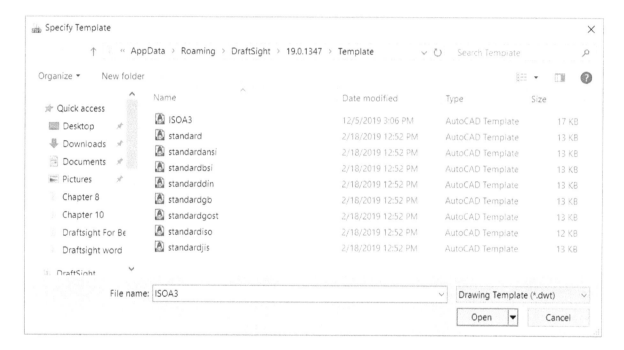

Changing the Color Scheme

DraftSight 2019 is available in two different color schemes: **Dark** and **Light**. You can change the color scheme by using the **Options** dialog. Click the right mouse button and select **Options** from the shortcut menu or click DraftSight menu on the top left corner of the **Main menu**.

On the **Options** dialog, click the **System Options** tab and expand **Display > Element Colors**. Now, select **Model background** from the **Element colors** list box. Next, select the color from the **Colors** drop-down. Click **Apply** and click **OK** on the **Options** dialog.

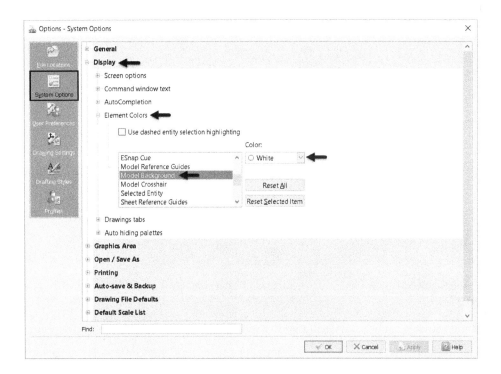

Help

Press **F1** or click the **Help** (?) icon located at the top right corner of the window to get help for any topic. On the Draftsight Help window, type-in a keyword in the search bar located on the left side; the topics are listed below it. Next, double-click on the desired topic in the list.

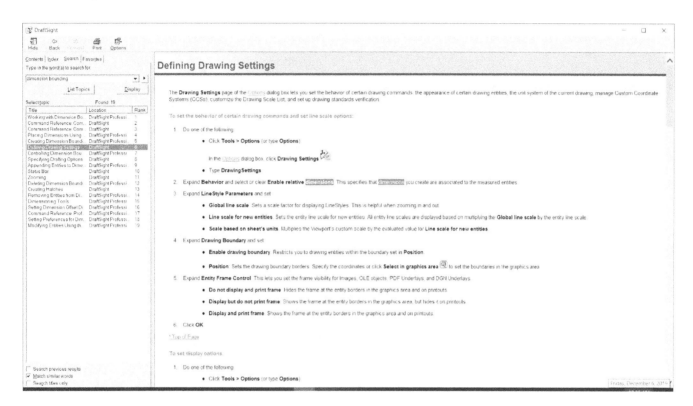

Chapter 2: Drawing Basics

In this chapter, you learn to do the following:

- **Draw lines, rectangles, circles, ellipses, arcs, polygons, and polylines**
- **Use the Erase, Undo and Redo tools**
- **Draw entities using the absolute coordinate points**
- **Draw entities using the relative coordinate points**
- **Draw entities using the tracking method**

Drawing Basics

This chapter teaches you to create simple drawings. You create these drawings using the basic drawing tools. These tools include **Line**, **Circle**, **Polyline**, and **Rectangle**, and they are available on the **Draw** group of the **Home** ribbon tab, as shown below. You can also activate these tools by typing them in the command window.

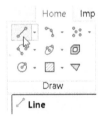

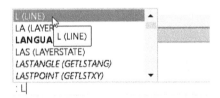

Drawing Lines

You can draw a line by specifying its start point and endpoint using the **Line** tool. However, there are various methods to specify the start and end of a line. These methods are explained in the following examples.

Example 1 (using the Cartesian Coordinate System)

In this example, you create lines by specifying points in the Cartesian coordinate system. In this system, you specify the points with respect to the origin (0, 0). You can specify a point by entering its X and Y coordinates separated by a comma, as shown in the figure below.

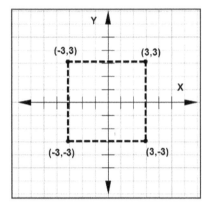

- Start DraftSight 2019 by clicking the **DraftSight 2019** icon on your desktop.
- Select the **Drafting and Annotation** workspace from the **Workspace** drop-down.
- Click the **New** icon on the Quick Access Ribbon.

- On the **Specify Template** dialog, select the **standardiso** template.

- Click the **Open** button.
- Click **View** tab > **Navigate** group >**Zoom Window** > **Zoom Fit** on the ribbon; the entire area in the graphics window is displayed.

- Make sure that the **Grid** [F7] icon is not turned ON on the status bar.

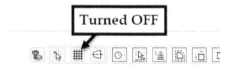

- Turn OFF the **QInput** icon on the status bar. You learn about **QInput** later in this chapter.

- To draw a line, click **Home > Draw > Line** on the Ribbon, or enter **LINE** or **L** in the command line.
- Type **50, 50** and press ENTER; the start point of the drawing is defined
- Type **150, 50,** and press ENTER.

- Type **150,100** and press ENTER.
- Type **50,100** and press ENTER.
- Type **C,** and press ENTER. This action creates a rectangle, as shown below.

- Click **Save** on the **Quick Access Ribbon**.

- Save as **Line-example1.dwg**.
- Close the file.

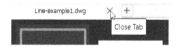

Example 2 (using Relative Coordinate system)

In this example, you draw lines by defining its endpoints in the relative coordinate system. In the relative coordinate system, you define the location of a point with respect to the previous point. For this purpose, the symbol, '@' is used before the point coordinates. This symbol means that the coordinate values are defined in relation to the previous point.

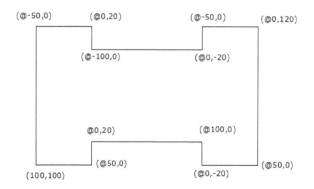

- Click the **New icon** on the **Quick Access Ribbon**.

- Select the **standardiso** template. Click **Open**.
- Type-in **Z** in the command line to activate the **ZOOM** command.
- Type B in the command line. Next, press Enter; the entire drawing is displayed in the graphics window.
- Turn OFF the **Grid** icon on the status bar.

- Turn OFF the **QInput** mode, if active.
- Click **Home > Draw > Line** on the ribbon, or enter **LINE** or **L** in the command line.
- Type **100,100** and press ENTER. The first point of the line is defined.
- Type **@50,0** and press ENTER.
- Type **@0,20** and press ENTER.
- Type **@100,0** and press ENTER.
- Type **@0,-20,** and press ENTER.
- Type **@50,0** and press ENTER.
- Type **@0,120** and press ENTER.
- Type **@-50,0** and press ENTER.
- Type **@0,-20,** and press ENTER.
- Type **@-100,0** and press ENTER.
- Type **@0,20** and press ENTER.
- Type **@-50,0** and press ENTER.
- Type **C** in the command line.
- Press Enter.
- Save the file as **Line-example2.dwg**.
- Close the file.

Example 3 (using the Polar Coordinate system)

In the polar coordinate system, you define the location of a point by entering two values: distance from the previous point and angle from the zero degrees. You enter the distance value along with the @ symbol and angle value with the < symbol. You have to make a note that DraftSight measures the angle in the anti-clockwise direction.

Drawing Task

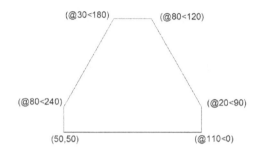

- Open a new file using the **standardiso.dwt** template.
- Click **View > Navigate > Zoom Bounds > Zoom Fit** on the ribbon.
- Turn OFF the **Grid** icon on the status bar.
- Turn OFF the **QInput** mode, if active.
- Click **Home > Draw > Line** on the Ribbon, or enter **LINE** or **L** in the command line.
- Type **50,50** and press **Enter** key.
- Type **@110<0** and press ENTER.
- Type **@20<90** and press ENTER.
- Type **@80<120** and press ENTER.
- Type **@30<180** and press ENTER.
- Type **@80<240** and press ENTER.
- Type **C** in the command line.
- Press Enter.
- Save the file as **Line-example3.dwg**.
- Close the file.

Example 4 (using Direct Distance Entry)

In the direct distance entry method, you draw a line by entering its distance and angle values. You use the **QInput** mode in this method.

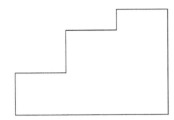

- Open a new file using the **standardiso.dwt** template.
- Turn OFF the **Grid** and **Snap** icons on the Status Bar.

- Click **View > Navigate > Zoom Window > Zoom Fit** on the ribbon.
- Activate the **QInput** icon on the Status Bar.

- Click **Home > Draw > Line** on the Ribbon, or enter **LINE** or **L** in the command line.
- Define the first point of the line by typing **50,50** and pressing ENTER.
- Move the pointer horizontally toward right and type in **150** in the length box.
- Press the TAB key and type **0** as the angle. Next, press ENTER.

- Move the pointer vertically upwards and type-in **100** as length.
- Press the TAB key and type **90** as the angle — next, press ENTER.
- Move the pointer horizontally toward left and type **50**.
- Press the TAB key and type **180** as the angle — next,

press ENTER.
- Move the pointer vertically downwards and type **20**.
- Press the TAB key and type **270** as the angle — next, press ENTER.
- Move the pointer horizontally toward left and type **50**.
- Press the TAB key and type **180** as the angle — next, press ENTER.
- Move the pointer vertically downwards and type **40**.
- Press the TAB key and type **270** as the angle — next, press ENTER.
- Move the pointer horizontally toward left and type **50**.
- Press the TAB key and type **180** as the angle — next, press ENTER.
- Type **Close** in the command line. Next, press Enter
- Save and close the file.

Erasing, Undoing and Redoing

- Draw the sketch similar to the one shown below using the **Line** tool.

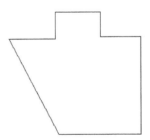

- Click **Home > Modify > Delete** on the Ribbon or Enter **DELETE** or **E** in the command line.

- Select the lines shown below and press ENTER. The lines are erased.

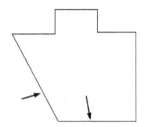

- Click the **Undo** button on the **Quick Access Ribbon**. The lines are restored.

- Click the **Redo** button on the **Quick Access Ribbon**. The lines are erased again.

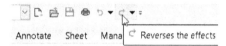

- Type **E** in the command line and press the SPACEBAR; the **DELETE** command is activated.

- Drag a selection window as shown below and press ENTER; the entities are erased.

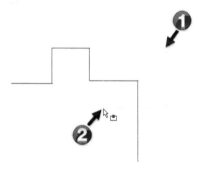

Drawing Circles

The tools in the **Circle** drop-down on the **Draw** group of the **Home** ribbon tab can be used to draw circles. You can also type-in the **CIRCLE** command in the command line and create circles. There are various methods to create circles. These methods are explained in the following examples.

Example 1(Circle)

In this example, you create a circle by specifying its center and radius value.

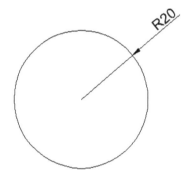

- Click **Home > Draw > Circle > Circle** on the Ribbon.

- Select an arbitrary point in the graphics window to specify the center point.

- Type 20 as the radius and press ENTER.

Example 2(Center, Diameter)

In this example, you create a circle by specifying its center and diameter value.

- Click **Home > Draw > Circle > Center, Diameter** on the ribbon. The message, "Specify diameter," appears in the command line.

- Pick a point in the graphics window, which is approximately horizontal to the previous circle.

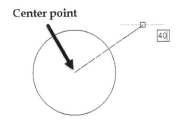

Center point

- Type 40 as the diameter and press ENTER; the circle is created.

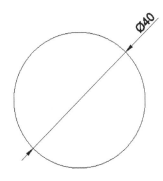

Ø40

Example 3(2-Point)

In this example, you create a circle by specifying two points. The first point is to specify the location of the circle, and the second defines the diameter.

- Right click on the **Esnap [F3]** icon on the status bar.
- Select the **Settings** option from the shortcut menu; the **Options – User Preferences** dialog appears.
- On the **Options – User Preferences** dialog, expand the **EntitySnaps** node. You learn about the EntitySnaps later in Chapter 3.
- Activate the **Center** option, if it is not already active.
- Click **OK** on the **Options – User Preferences** dialog.

Now, you create a circle by selecting the center points of the previous circles.

- Click **Home > Draw > Circle > 2-Point** on the Ribbon. The message, "Specify diameter start point,"

appears in the command line.

- Select the center point of the left side circle; the message, "Specify diameter end point," appears in the command line.
- Select the center point of the right-side circle; the circle is created, as shown below.

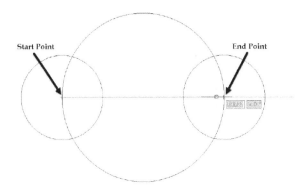

Start Point   End Point

Example 4(3-Point)

In this example, you create a circle by specifying three points. The circle passes through these three points.

- Open a new file.
- Use the **Line** tool and create the drawing shown in the figure below. The coordinate points are also given in the figure.

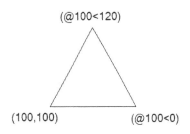

(@100<120)

(100,100)   (@100<0)

- Click **Home > Draw > Circle > 3-Point** on the Ribbon.
- Select the three vertices of the triangle; a circle is created, passing through the selected points.

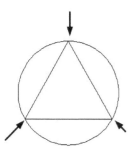

Example 5 (Tangent, Tangent, Radius)

In this example, you create a circle by selecting two objects and then specifying the radius of the circle. A circle tangent to the selected objects is created.

- Select the circle passing through the three vertices of the triangle; the radius and diameter values of the circle are displayed in the Properties palette, as shown.

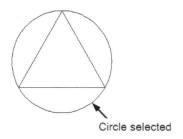

Circle selected

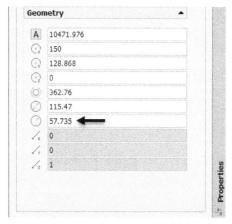

- Click **Manage > Utilities > Smart Calculator** on the ribbon; the **Calculator** appears.
- Type-in **57.7350** in the **expression** box on the **Calculator**.

- Click the / button and then the **2** button on the **Number Pad**.
- Click the = button; the value **28.8675** is displayed in the expression box.

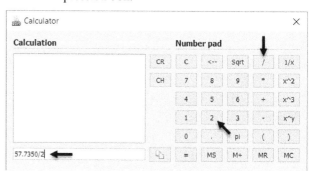

- Click the **Copy** icon next to the expression box; the value **28.8675** is copied.
- Click the **Close** button on the **Calculator**.
- Click **Home > Draw > Circle > Tangent Tangent Radius** on the ribbon. The message, "Select first tangency," appears in the command line.
- Select the horizontal line of the triangle; the message, "Specify second tangency:" appears in the command line.
- Select any one of the inclined lines; the message, "Specify radius," appears in the command line.
- Click in the command line and press Ctrl+V on your keyboard; the value **28.8675** is pasted in the command line.

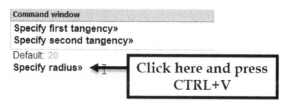

- Press ENTER to specify the radius; the circle is created, touching all three sides of the triangle.

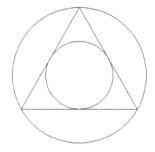

- Save and close the file.

Example 6 (Tangent, Tangent, Tangent)

In this example, you create a circle by selecting three objects to which it will be tangent.

- Click the **Open** button on the **Quick Access Ribbon**; the **Open** dialog appears.

- Browse to the location of the **Line-example3.dwg** file and double-click on it; the file is opened.
- Click **Home > Draw > Circle > Tangent, Tangent, Tangent** on the ribbon.
- Select the bottom horizontal line of the drawing.
- Select the two inclined lines; a circle tangent to the selected lines is created.

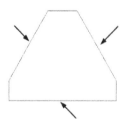 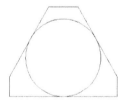

- Save and close the file.

Drawing Arcs

An arc is a portion of a circle. The total angle of an arc is always less than 360 degrees, whereas the total angle of a

circle is 360 degrees. DraftSight provides you with eleven ways to draw an arc. You can draw arcs in different ways by using the tools available in the **Arcs** drop-down of the **Draw** ribbon. The usage of these tools depends on your requirements. Some methods to create arcs are explained in the following examples.

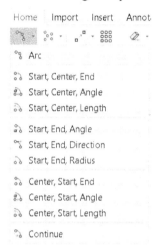

Example 1 (Arc)

In this example, you create an arc by specifying three points. The arc passes through these points.

- Open the **Line-example1.dwg** file.
- Click the **Home > Draw > Multiple Points** command on the Ribbon.

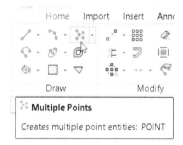

- Type 100,120 in the command line and press ENTER. This places a point above the rectangle.

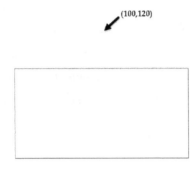

(100,120)

- Right click on the **Esnap [F3]** icon on the status bar.
- Select the **Settings** option from the shortcut menu; the **Options – User Preferences** dialog appears.
- On the **Options – User Preferences** dialog, expand the **EntitySnaps** node.
- Activate the **Node** option, if it is not already active.
- Click **OK** on the **Options – User Preferences** dialog.
- Click **Home > Draw > Arc > Arc** on the Ribbon. The message, "Specify start point," appears in the command line.
- Select the top left corner of the rectangle.
- Select the point located above the rectangle.
- Select the top right corner of the rectangle; the three-point arc is created.

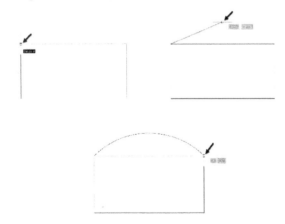

Example 2 (Start, Center, End)

In this example, you draw an arc by specifying its start, center, and endpoints. The first two points define the

radius of the arc, and the third point defines its included angle.

- Click **Home > Draw > Arc > Start, Center, End** on the Ribbon. The message, "Specify start point," appears in the command line.

The included angle of the arc is measured in the counter-clockwise direction. Press and hold the Ctrl key if you want to reverse the direction.

- Pick an arbitrary point in the graphics window to define the start point of an arc. The message, "Specify center point," appears.
- Pick a point to define the radius of the circle. You can also type in the radius value and press ENTER; the message, "Specify end point," appears. Notice that, as you move the pointer, the included angle of the arc changes.
- Pick a point to define the included angle of the arc. You can also type the angle value and press ENTER.

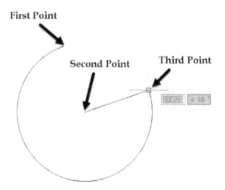

First Point

Second Point   Third Point

Example 3 (Start, End, Direction)

- Use the **Line** tool and create the drawing shown in the figure below. The dimensions are also given in the figure. (Use any one of the procedures given in the **Drawing Lines** section)

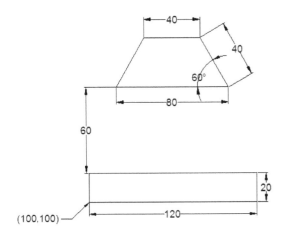

- Click **Home > Draw > Arc > Start, End, Direction** on the Ribbon.

- Specify the start and endpoints of the arc, as shown in the figure.

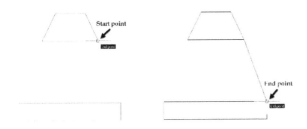

- Move the pointer vertically downward and click to specify the direction.

- Likewise, create another arc.

Drawing Polylines

A Polyline is a single object that consists of line segments and arcs. It is more versatile than a line, as you can assign a width to it. In the following example, you create a closed polyline.

Example 1

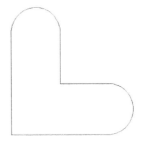

- Activate the **Ortho** icon on the Status Bar.

- Click **Home > Draw > Polyline** on the Ribbon or enter **PLINE** or **PL** in the command line; the message, "Specify start point:" appears in the command line.

- Select an arbitrary point in the graphics window.

- Move the pointer horizontally toward right and type 100 — next, press ENTER.

- Type **A** or **Arc** in the command line and press ENTER.

- Move the pointer vertically upward and type **50** — next, press ENTER.

- Type **L** or **Line** in the command line and press ENTER.

- Move the pointer horizontally toward left and type **50** — next, press ENTER.

- Move the pointer vertically upward and type **50** — next, press ENTER.

- Type **A** or **Arc** in the command line and press ENTER.

- Move the pointer horizontally toward left and type **50** — next, press ENTER.

- Type **CL** or **CLose** in the command line and press ENTER.

The polyline created is a single object. As a result, the whole sketch is selected when you select a line segment from the sketch.

Drawing Rectangles

A rectangle is a single four-sided object. You can create a rectangle by just specifying its two diagonal corners. However, there are various methods to create a rectangle. These methods are explained in the following examples.

Example 1

In this example, you create a rectangle by specifying its corner points.

- Open a new file.
- Click **Home > Draw > Polygon** drop-down > **Rectangle** on the Ribbon or enter **RECTANG** or **REC** in the command line; the message, "Specify start corner," appears in the command line.
- Pick an arbitrary point in the graphics window; the message "Specify opposite corner" appears in the command line.
- Move the pointer diagonally toward the right and click to create a rectangle.

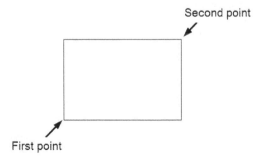

Example 2

In this example, you create a rectangle by specifying its length and width.

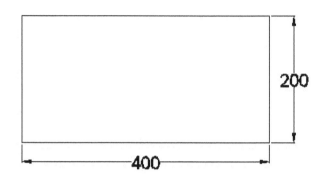

- Click **Home > Draw > Polygon** drop-down > **Rectangle** on the Ribbon, or enter **RECTANG** or **REC** in the command line.
- Specify the start corner of the rectangle by picking an arbitrary point in the graphics window.
- Follow the prompt sequence given next:
  Options: Area, Dimensions, Rotation or Specify opposite corner >>
  Type **D** or **Dimensions** in the command line and press ENTER
  Specify horizontal dimension >> Type **400** and press ENTER.
  Specify vertical dimension >> Type **200** and press ENTER.

Example 3

In this example, you create a rectangle by specifying its area and width.

- Click **Home > Draw > Polygon** drop-down > **Rectangle** on the Ribbon, or enter **RECTANG** or

REC in the command line.

- Specify the start corner of the rectangle by picking an arbitrary point.

- Follow the prompt sequence given next:

Options: Area, Dimensions, Rotation or Specify opposite corner >> Type **A** or **Area** in the command line and press ENTER

Specify total area: Type **20000** and press ENTER.

Options: Horizontal or Vertical

Specify known dimension >> Type **V** or **Vertical** for the vertical dimension. You can also type **H** or **Horizontal** for the horizontal dimension in the command line. Next, press ENTER.

Enter horizontal dimension >> (This appears, only if you have entered **Horizontal**) Type **100** and press ENTER; the length is calculated automatically.

Example 4

In this example, you create a rectangle with chamfered corners.

- Click **Home > Draw > Polyline** drop-down > **Rectangle** on the Ribbon, or enter **RECTANG** or **REC** in the command line.

- Follow the prompt sequence given next:

Options: 3Corner, 3Point center, CEnter, COrner, Chamfer, Elevation, Fillet, Parallelogram, Thickness, line Width or

Specify start corner >>

Type **C** or **Chamfer** option in the command line and press ENTER.

Specify first chamfer length >> Type **20** and press ENTER.

Specify second chamfer length >> Type **20** and press ENTER.

Options: 3Corner, 3Point center, CEnter, COrner, Chamfer, Elevation, Fillet, Parallelogram, Thickness, line Width or

Specify start corner >>

Click at an arbitrary point in the graphics window to specify the first corner.

Options: Area, Dimensions, Rotation or

Specify opposite corner >>

Move the pointer diagonally toward the right and click to specify the second corner.

Example 5

In this example, you create a rectangle with rounded corners.

- Click **Home > Draw > Polyline** drop-down > **Rectangle** on the Ribbon, or enter **RECTANG** or **REC** in the command line.

- Follow the prompt sequence given next:

Options: 3Corner, 3Point center, CEnter, COrner, Chamfer, Elevation, Fillet, Parallelogram, Thickness, line Width or

Specify start corner >>

Type **F** or **Fillet** in the command line and press ENTER.

Specify fillet radius >> Type **50** and press ENTER.

Options: 3Corner, 3Point center, CEnter, COrner, Chamfer, Elevation, Fillet, Parallelogram, Thickness, line Width or

Specify Start corner >>

Click at an arbitrary point in the graphics window to specify the first corner.

Options: Area, Dimensions, Rotation or

Specify opposite corner >> Move the pointer diagonally toward the right and click to specify the second corner.

Example 6

In this example, you create an inclined rectangle.

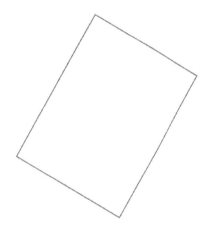

- Click **Home > Draw > Polyline** drop-down > **Rectangle** on the Ribbon, or enter **RECTANG** or **REC** in the command line.
- Specify the first corner of the rectangle by picking an arbitrary point.
- Follow the prompt sequence given next:

Options: Area, Dimensions, Rotation or

Specify opposite corner>>

Type **F or Fillet** in the command line and press ENTER.

Specify fillet radius >> Type **0** and press ENTER.

Type **R** or **Rotation** in the command line and press ENTER.

Pick points or

Specify rotation >>

Type **60** and press ENTER.

Options: Area, Dimensions, Rotation or

Specify opposite corner >>

Type **D** or **Dimensions** in the command line and press Enter.

Specify horizontal dimension >> Type 400 and press ENTER.

Specify vertical dimension >> Type 300 and press ENTER.

Drawing Polygons

A Polygon is a single object having many sides ranging from 3 to 1024. In DraftSight, you can create regular polygons having sided with equal length. There are two methods to create a polygon. These methods are explained in the following examples.

Example 1

In this example, you create a polygon by specifying the number of sides and then specifying the length of one side.

- Click **Home > Draw > Polyline** drop-down > **Polygon** on the ribbon.

- Follow the prompt sequence given next:

Default: 4

Specify number of sides >> Type **5** and press ENTER.

Options: Side length or

Specify center point >> Type **S** or Side length in the command line and press ENTER.

Specify start point >> Select an arbitrary point.

Specify side length >> Move the pointer horizontally and type **20** — next, press ENTER.

Example 2

In this example, you create a polygon by specifying the number of sides and drawing an imaginary circle (inscribed circle). The polygon is created with its corners located on the imaginary circle. You can also create a polygon with the circumscribed circle. A circumscribed circle is an imaginary circle which is tangent to all the sides of a polygon.

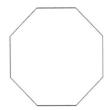

- Type **POL** in the command line and press ENTER; the **POLYGON** command is activated.
- Follow the prompt sequence given next:

Default: 5

Specify number of sides >> Type **8** and press ENTER.

Options: Side length or

Specify center point >> Select an arbitrary point

Default: COrner

Options: COrner or Side

Specify distance option >> Type **CO** or **COrner** in the command line.

Specify distance >> Type **20** and press ENTER; a polygon is created with its corners touching the imaginary circle.

Drawing Splines

Splines are non-uniform curves, which are used to create irregular shapes. In DraftSight, you can create splines by using two methods: **Spline Fit** and **Spline CV**. These methods are explained in the following examples:

Example 1: (Spline Fit)

In this example, you create a spline using the **Spline Fit** method. In this method, you need to specify various points in the graphics window. The spline is created, passing through the specified points.

- Start a new drawing file.
- Use the **Line** tool and create a sketch similar to the one shown below.

- Click **Home > Draw > Ellipse** drop-down **> Spline** on the Ribbon.

  the message, "Specify start fit point," appears in the command line.

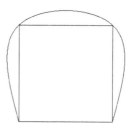

- Select the lower-left corner of the sketch; the message, "Specify next fit point," appears in the command line.
- Select the top-left corner point of the sketch; the message, "Options: Close, Fit tolerance, Enter for start tangency or Specify next fit point," appears in the command line.
- Similarly, select the top-right and lower-right corners; a spline is attached to the pointer.

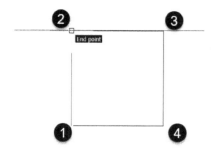

- Right-click and select Enter; the message, "Specify Start tangency," appears in the command line.

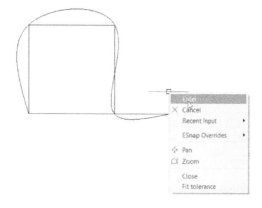

- Type 0 and press Enter; the message, "Specify End tangency," appears in the command line.
- Type 0 and press Enter.

Drawing Ellipses

Ellipses are also non-uniform curves, but they have a regular shape. They are splines created in a regular closed manner. In DraftSight, you can draw an ellipse in three different ways by using the tools available in the **Ellipse** drop-down on the **Draw group** of the Ribbon. The three different ways to draw ellipses are explained in the following examples.

Example 1 (Center)

In this example, you draw an ellipse by specifying three points. The first point defines the center of the ellipse. The second and third points define the two axes of the ellipse.

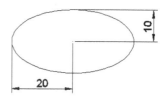

- Click **Home > Draw > Ellipse** drop-down **> Center** on the Ribbon; the message, "Specify center point>>" appears in the command line.
- Select an arbitrary point in the graphics window; the message, "Specify axis end point>>," appears in the command line.
- Move the pointer horizontally and type **20**. Next, press ENTER; the message, "Options: Rotation or Specify other axis end point>>" appears in the command line.

- Type **10** and press ENTER; the ellipse is created.

Example 2 (Axis, End)

In this example, you draw an ellipse by specifying three points. The first two points define the location and length of the first axis. The third point defines the second axis of the ellipse.

- Activate the **QInput** icon on the status bar, if it is not active.
- Deactivate the **Ortho** icon on the status bar.
- Click **Home** > **Draw** > **Ellipse** drop-down > **Axis, End** on the Ribbon.
- Select an arbitrary point to specify the start axis.
- Type **50** as the length of the first axis and press TAB.
- Type **60** as angle and press ENTER.
- Type **10** as the radius of the other axis and press ENTER; the ellipse is created inclined at a 60-degree angle.

Example 3 (Elliptical Arc)

In this example, you draw an elliptical arc. To draw an elliptical arc, first, you need to define the location and length of the first axis. Next, define the radius of the second axis; an ellipse is displayed. Now, you need to define the start angle of the elliptical arc. The start angle can be any angle between 0 and 360. After defining the start angle, you need to specify the end angle of the

elliptical arc.

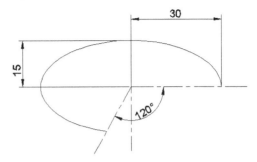

- Turn on the **Ortho icon** on the Status bar.

- Click **Home** > **Draw** > **Ellipse** drop-down > **Elliptical Arc** on the Ribbon.
- Select an arbitrary point to specify an axis endpoint.
- Move the pointer horizontally toward left and type **60**. Next, press ENTER to specify the axis start point.
- Move the pointer upward and type **15**. Next, press ENTER to specify the other axis endpoint.
- Type **0** and press ENTER to specify the start angle.
- Type **240** and press ENTER to specify the end angle.

Exercises

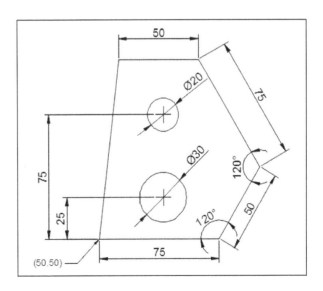

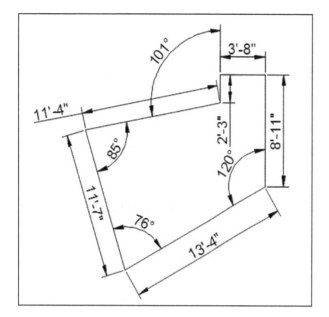

Chapter 3: Drawing Aids

In this chapter, you learn to do the following:

- **Use Grid and Snap**
- **Use Ortho Mode and Polar Tracking**
- **Use ESnaps and tracking**
- **Create Layers and assign properties to it**
- **Zoom and Pan drawings**

Drawing Aids

This chapter teaches you to define the drafting settings, which assist you in creating a drawing in DraftSight easily. Most drafting settings can be turned on or off from the status bar. You can also access additional drafting settings in the **Options** dialog.

Grid and Snap Settings

Grid is the basic drawing setting. It makes the graphics window appear like a graph paper. You can turn ON the grid display by clicking the **Grid** icon on the status bar or just pressing **F7** on the keyboard.

Snap is used for drawing objects by using the intersection points of the grid lines. When you turn the **Snap** icon ON, you can select only grid points. In the following example, you learn to set the grid and snap settings.

Example:

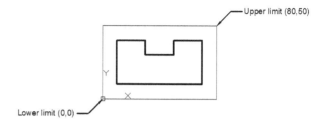

- Click **Application Menu > New**; the **Specify Template** dialog appears.
- Select the **standardiso** template. Click **Open**.
- On the Status bar, click the right mouse button on the **Snap** icon and select **Settings**. The **Options** dialog appears with the **Snap Settings** node expanded.
- Check the **Enable Snap (F9)** option on the dialog under **Snap Settings**.
- Check the **Match Grid spacing** option.
- Set **Horizontal Snap spacing** to **10** and **Vertical Snap spacing** to **10** under the **Spacing** section.
- Check the **Match horizontal spacing** option.

- Go to the **Display** node and expand the **Grid Settings** sub-node.
- Make sure that the **Enable Grid (F7)** option is selected.

- Click **OK** on the dialog.

Setting the Limits of a drawing

You can set the limits of a drawing by defining its lower-left and top-right corners. By setting Limits of a drawing, you define the size of the drawing area. In DraftSight, limits are set to some default values. However, you can redefine the limits to change the drawing area as per your requirement.

- Type **Limits** or **DrawingBounds** in the command line and press ENTER.

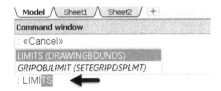

- Type **0,0** and press ENTER to define the lower-left corner.
  Now, you need to define the upper right corner.
- Type **80,50** and press ENTER key.
- On the **View** tab, click the **Navigate** group > **Zoom Window** drop-down > **Zoom Fit**; the graphics window is zoomed to the limits.

Setting the Lineweight

Line weight is the thickness of the objects that you draw. In DraftSight, there is a default lineweight assigned to objects. However, you can set a new lineweight. The method to set the lineweight is explained below.

- On the **Home** tab, click the **Lineweight** ☰ icon on the **Properties** panel; The **Options** dialog appears with **Lineweight: By Layer** section expanded.
- On the dialog, select **0.40** mm from the **Default weight** drop-down.
- Check the **Display weight in graphics area** option under the **Line weight: By Layer** section.

- Click **OK** on the dialog.
- Type L in the command line and press ENTER.
- Type 10,10 and press ENTER to define the first point.
- Move the pointer horizontally toward the right and click on the sixth grid point from the first point.

- Move the pointer vertically upwards and select the third grid point from the second point.

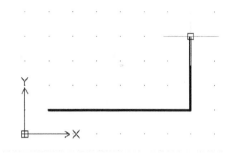

- Move the pointer horizontally toward the left and select the second grid point from the previous point.

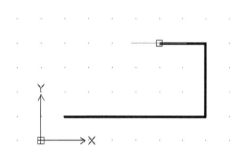

- Move the pointer vertically downwards and select the grid point next to the previous point.

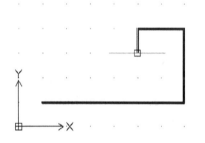

- Move the pointer horizontally toward the left and select the second grid point from the previous point.

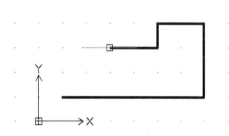

- Move the pointer vertically upwards and select the grid point next to the previous point.

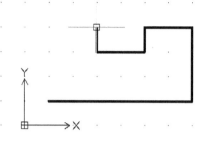

- Move the pointer horizontally toward the left and select the second grid point from the previous point.

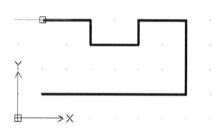

- Type **C** or **Close** in the command line.

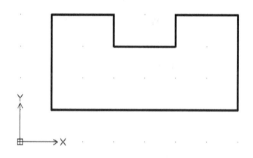

- Save and close the file.

Using Ortho and Polar

The **Ortho** option is used to draw orthogonal (horizontal or vertical) lines. Polar is used to constrain the lines to angular increments. In the following example, you create a drawing with the help of Ortho and Polar options.

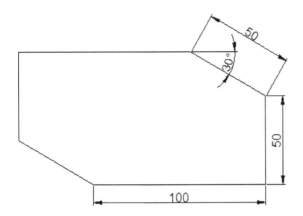

- Open a new DraftSight file.

- Deactivate the **Grid** and **Snap** icons on the status bar.

- On the Status bar, right-click on the **Polar** icon and select **Settings**.

- On the **Options** dialog, set the **Incremental angles for Polar guide display** to **30**.

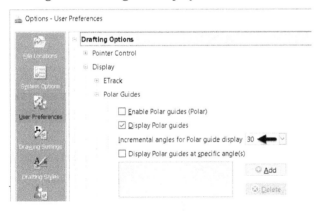

- Click **OK**.

- Activate the **Ortho Mode** icon on the status bar.

- Click **View** tab > **Navigate** panel > **Zoom Window** > **Zoom Fit** on the Ribbon.

- Click the **Line** button on the **Draw** panel.

- Select an arbitrary point to define the starting point.

- Move the pointer toward the right, type 100, and press ENTER; notice that a horizontal line is created.

- Move the pointer upwards, type 50, and press ENTER; notice that a vertical line is created.

- Activate the **Polar** icon on the status bar.

- Rotate the pointer and notice that a tracking line is displayed at every thirty-degree interval.

- Stop the pointer at the angle interval, as shown.

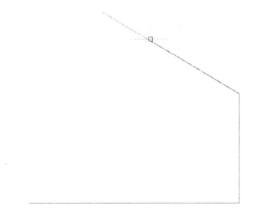

- Type 50 and press ENTER.

- Move the pointer toward left, and then stop when a track line is displayed horizontally.

- Type 100, and press ENTER.

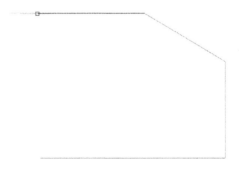

- Move the pointer vertically downward and notice that a vertical track line is displayed.

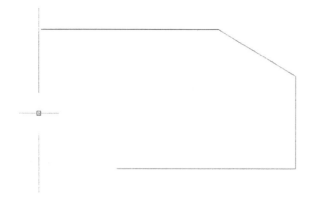

- Type 50 and press ENTER.
- Type **C** or **Close** in the command line.

Using Layers

Layers are like a group of transparent sheets that are combined into a complete drawing. The figure below displays a drawing consisting of object lines and dimension lines. In this example, the object lines are created on the 'Object' layer, and dimensions are created on the layer called 'Dimension.' You can easily turn-off the 'Dimension' layer for a clearer view of the object lines.

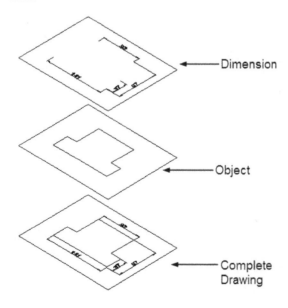

Layer Manager

The **Layer Manager** is used to create and manage layers. To open **Layer Manager**, click **Home > Layers > Layer Manager** on the ribbon or enter **LA** in the command line.

The components of the **Layer Manager** are shown below. The **List** section is the main body of the **Layer Manager**. It lists the individual layers that currently exist in the drawing.

The **List** section contains various properties. You can set layer properties and perform various operations in the **List** section. A brief explanation of each layer property is given below.

Status – Shows an arrow when a layer is set to current.
Name - Shows the name of the layer.
Show – Used to turn on/off the visibility of a layer. When a layer is turned on, it shows a green dot. When you turn off a layer, it shows a grey dot.
Frozen – It is used to freeze the objects of a layer so that they cannot be modified. Also, the visibility of the object is turned off.
Lock- It is used to lock the layer so that the objects on it cannot be modified.
LineColor – It is used to assign a color to the layer.
LineStyle – It is used to assign a linetype to the layer.

LineWeight – It is used to define the lineweight (thickness) of objects on the layer.

Transparency – It is used to define the transparency of the layer. You set a transparency level from 0 to 90 for all objects on a layer.

PrintStyle – It is used to override the settings such as color, linetype, and lineweight while plotting a drawing.

Print – It is used to control which layer will be printed.

New ViewPoint – It is used to create and freeze a layer in any new viewport.

Description – It is used to enter a detailed description of the layer.

Creating a New Layer

You can create a new layer by using anyone of the following methods:

1. Click the **New Layer** button on the **Layer Manager**; a new layer with the name 'Layer1' appears in **Name** field. Next, enter the name of the layer in the **Name** field.

2. Right-click in the **Name** field and select **New Layer** from the shortcut menu.

Making a layer current

If you want to draw objects on a particular layer, then you have to make it current. You can make a layer current using the methods listed below.

1. Select the layer from the List view and click the **Activate** button on the **Layer Manager**.

2. Double-click on the **Status** field of the layer.

3. Right-click on the layer and select **Activate Layer**.

4. Select the layer from the **Layer** drop-down of the **Layer** panel.

5. You can also click the **Activate Layer** icon on the **Layer** panel to make the layer current. Next, select an object; the layer related to the selected object becomes current.

Deleting a Layer

You can delete a layer by using anyone of the following methods:

1. Click the **Delete Layer** button or press ALT+D.

2. Right-click in the **Name** field and select **Delete Layer** from the shortcut menu.

3. Click the down-arrow next to the **Activate Layer** icon and select the **Delete Layer** option. Next, select an object; the layer related to the selected object is deleted.

You learn more about layers in later chapters. You can find an example related to layers in the **Offset** tool section of chapter 4.

Using ESnaps

ESnaps are important settings that improve your performance and accuracy while creating a drawing. They allow you to select key points of objects while creating a drawing. You can activate the required ESnap by using the **Esnap** shortcut menu. Press and hold the SHIFT key and right-click to display this shortcut menu.

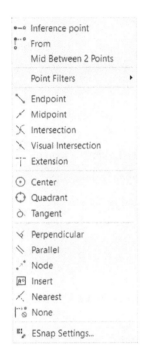

The functions of various ESnaps are explained next.

Endpoint: Snaps to the endpoints of lines and arcs. For example, press and hold the SHIFT key and right click to display the ESnap menu. Next, select the Endpoint option from the Esnap menu. Click on a line or arc; the endpoint of the object is selected.

Midpoint: Snaps to the midpoint of a line.

Intersection: Snaps to the intersections of objects.

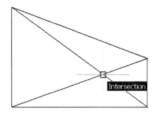

Visual Intersection: Snaps to the projected intersection of two objects in 3D space.

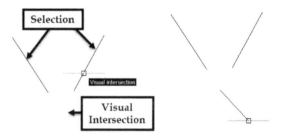

Center: Snaps to the centers of circles and arcs.

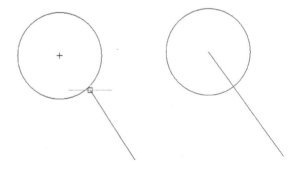

Quadrant: Snaps to four key points located on a circle.

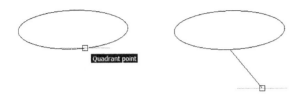

Tangent: Snaps to the tangent points of arcs and circles.

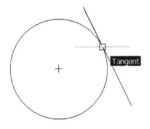

Perpendicular: Snaps to a perpendicular location on an object.

Node: Snaps to points of dimension lines, text objects, and dimension text.

Insert: Snaps to the insertion point of blocks, shapes, and text.

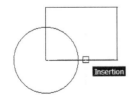

Nearest: Snaps to the nearest point found along with any object.

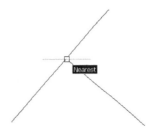

From: Locates a point at a specified distance and direction from a selected reference point.

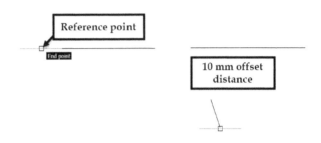

Midpoint Between 2 Points: Snaps to the middle point of two selected points.

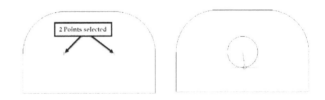

Running ESnaps

Previously, you have learned to select Esnaps from the shortcut menu. However, you can make ESnaps available continuously instead of selecting them every time. You can do this by using the ESnap **Settings**. To use the Esnaps, right-click the **ESnaps** icon on the status bar and select **Settings**. Select the required snap by selecting checkboxes on the **Options** dialog. Also, make sure that the **Enable EntitySnaps(ESnaps)** is checked.

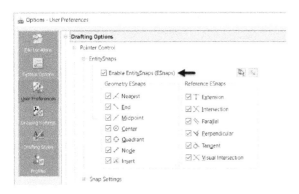

Cycling through ESnap

After setting the Running ESnap settings, DraftSight displays object snaps depending on the shape of the object. However, you can cycle through the object snaps by pressing the TAB key. In the following example, you learn to cycle through different ESnaps.

Example:

- Right-click on the **ESnap** icon and select the **Settings** option; the **Options** dialog appears. The EntitySnaps section is expanded. Click **Select All** button and click the **OK** button.
- Draw the objects, as shown below.

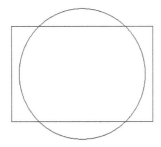

- Click the **Circle** button on the **Draw** panel.
- Place the pointer on the drawing. Press the TAB key; notice that the object snaps change.

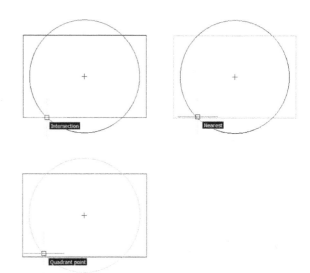

- Click when the **Center** snap is displayed and draw a circle.

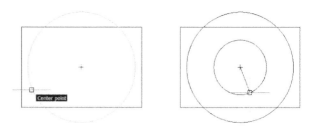

Using ETrack

ETracking is the movement of the pointer along the trace lines originating from the keypoints of objects. ETrack works only when the **ESnap** mode is turned on. In the following example, you learn to use ETrack for creating objects.

Example:

- Select the **ETrack** button from the Status bar.

(OR)

- Open the **Options** dialog and click **User Preferences** > **Drafting Options.**

- Expand the **Display** section and then expand the **ETrack** section.
- Make sure that the **Enable EntityTracking (ETrack)** is checked.

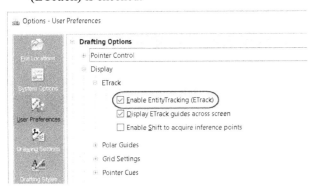

- Click **OK**.
- Use the **Line** tool and draw the objects, as shown below.

- Press the ENTER key twice to start drawing lines from the last point.
- Move the pointer and place it on the endpoint of the lower horizontal line.

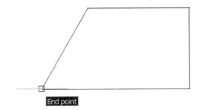

- Move the pointer vertically upward; notice the trace line, as shown below.

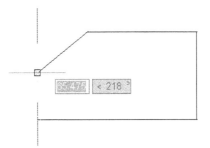

- Click on the trace line to create an inclined line.
- Snap the pointer to the endpoint of the lower horizontal line and click.

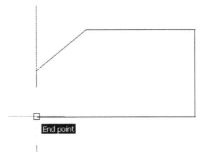

- Right-click and select **Enter**.
- Click the **Circle** button on the **Draw** group of the **Home** ribbon tab.
- Place the pointer over the lower endpoint of the inclined line and move horizontally; notice that a trace line is displayed.
- Place the pointer on the midpoint of the lower horizontal line; a vertical trace line is displayed from the midpoint of the horizontal line, as shown below.

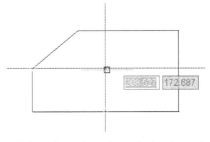

- Click at the point where the horizontal and vertical trace lines intersect. Next, create a circle, as shown below.

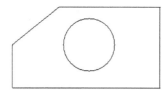

Using Zoom tools

Using the zoom tools, you can magnify or reduce a drawing. You can use these tools to view the minute details of a very complicated drawing. The Zoom tools can be accessed from the Command line, and Ribbon.

Zooming with the Mouse Wheel

Zooming using the mouse wheel is one of the easiest methods.

- Roll the mouse wheel forward to zoom into a drawing.
- Roll the mouse wheel backward to zoom out of the drawing.
- Press the mouse wheel and drag the mouse to pan the drawing.

Using Zoom Fit

Using the **Zoom Fit** tool, you can zoom to the extents of the largest object in a drawing.

- Click **View > Navigate > Zoom Window** drop-down > **Zoom Fit** on the ribbon.

- You can also double-click on the mouse wheel to zoom to extents.

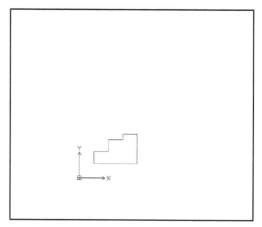

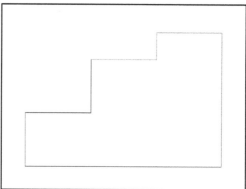

Using Zoom-Window

Using the **Zoom-Window** tool, you can define the area to be magnified by selecting two points representing a rectangle.

- Click **View > Navigate > Zoom Window** on the ribbon.
- Specify the first point of the zoom window, as shown.
- Move the pointer diagonally toward the right, and then specify the second point, as shown. The area inside the window is magnified.

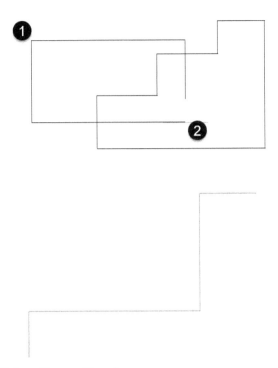

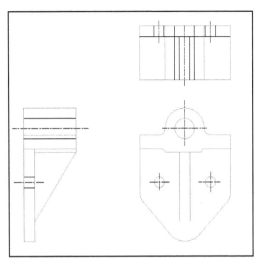

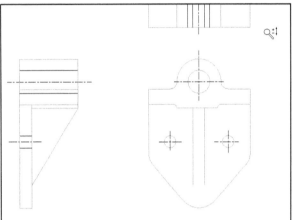

Using Zoom-Previous

After magnifying a small area of the drawing, you can use the **Zoom-Previous** tool to return to the previous display.

- Click **View > Navigate > Zoom window** drop-down > **Zoom Previous** on the ribbon.

Using Dynamic Zoom

Using the **Zoom** tool, you can zoom in or zoom out of a drawing dynamically.

- Click **View > Navigate > Dynamic Zoom** on the ribbon; the pointer is changed to a magnifying glass with plus and minus symbols.
- Press and hold the left mouse button and drag the mouse forward to zoom into the drawing.

- Drag the mouse backward to zoom out of the drawing.

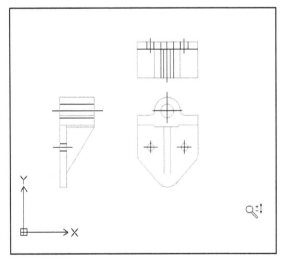

Using Zoom Factor

Using the **Zoom Factor** tool, you can zoom in or zoom out of a drawing by entering zoom scale factors directly from your keyboard.

- Click **View** > **Navigate** > **Zoom Window** drop-down > **Zoom Factor** on the ribbon. The message, nX or nXP "**Specify scale factor >>**" appears in the command line.
- Enter the scale factor 0.25 to scale the drawing to 25% of the full view.
- Enter the scale factor 0.25X to scale the drawing to 25% of the current view.
- Enter the scale factor 0.25XP to scale the drawing to 25% of the paper space.

Using Zoom Center

Using the **Zoom Center** tool, you can zoom to an area of the drawing based on a center point and magnification value.

- Click **View** > **Navigate** > **Zoom Window** drop-down > **Zoom Center** on the ribbon; the message, "**Specify Center point,**" appears in the command line.
- Select a point in the drawing to which you want to zoom in; the message, "**Specify magnification or height,**" appears in the command line.

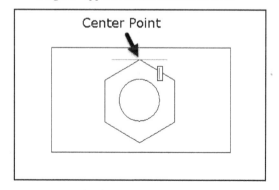

- Enter 10 in the command line to magnify the location of the point by ten times.

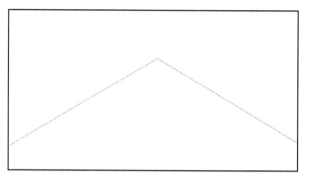

Using Zoom Selected

Using the **Zoom Selected** tool, you can magnify a portion of the drawing by selecting one or more objects.

- Click **View** > **Navigate** > **Zoom Window** drop-down > **Zoom Selected** on the ribbon.
- Select one or more objects from the drawing and press ENTER; the objects are magnified.

Using Zoom In

Using the **Zoom In** tool, you can magnify the drawing by a scale factor of 2.

- Click **View** > **Navigate** > **Zoom Window** drop-down > **Zoom In** on the ribbon; the drawing is magnified to double.

Using Zoom Out

The **Zoom Out** tool is used to de-magnify the display screen by a scale factor of 0.5.

Dynamic Panning

After zooming into a drawing, you may want to view an area that is outside the current display. You can do this by using the **Dynamic Pan** tool.

- Click **Dynamic Pan** on the ribbon.

- Press and hold the left mouse button and drag the mouse; a new area of the drawing, which is outside the current view, is displayed.

Exercises

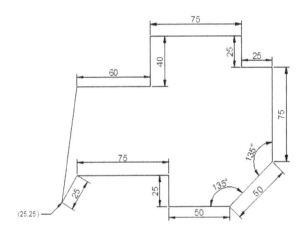

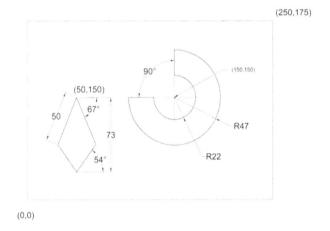

Chapter 4: Editing Tools

In this chapter, you learn the following tools:

- The **Move** tool
- The **Copy** tool
- The **Rotate** tool
- The **Scale** tool
- The **Trim** tool
- The **Extend** tool
- The **Fillet** tool
- The **Chamfer** tool
- The **Mirror** tool
- The **Explode** tool
- The **Stretch** tool
- The **Polar Array** tool
- The **Offset** tool
- The **Path Array** tool
- The **Rectangular Array** tool

Editing Tools

In previous chapters, you have learned to create some simple drawings using the basic drawing tools. However, to create complex drawings, you may perform various editing operations. The tools to perform the editing operations are available in the **Modify** panel on the **Home** ribbon. Using these editing tools, you can modify existing objects or use existing objects to create new or similar objects.

The Move tool

The **Move** tool moves a selected object(s) from one location to a new location without changing its orientation. To move objects, you must activate this tool and select the objects from the graphics window. After selecting the objects, you must define the 'base point' and the 'destination point.'

Example:

- Create the drawing, as shown below.

- Click **Home > Modify > Copy** drop-down > **Move** on the ribbon, or enter **M** in the command line.
- Click on the circle located at the right side, and then right-click to accept the selection.
- Select the center of the circle as the 'from point.'
- Make sure that the **Ortho** ⊟ icon is activated.
- Move the pointer toward the right and pick a point, as shown below. The circle is moved to the new location.

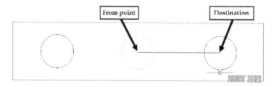

The Copy tool

The **Copy** tool is used to copy objects and place them at a required location. This tool is similar to the **Move** tool, except that the object remains at its original position, and a copy of it is placed at the new location.

Example:

- Draw two circles of 80 mm and 140 mm diameter, respectively.

- Click **Home > Modify > Copy** on the ribbon or enter **CO** in the command line.
- Select the two circles and then right-click to accept the selection.
- Select the center of the circle as the base point.
- Make sure that the **Ortho** icon is activated.
- Move the pointer toward the right.
- Type 200, and press ENTER.
- Enter **E** or **Exit** in the command line. A copy of the circles is created at the new location.

The Rotate tool

The **Rotate** tool rotates an object or a group of objects about a base point. Activate this tool and select the objects from the graphics window. After selecting objects, you must define the 'base point' and the angle of rotation. The object(s) is rotated about the base point.

- Click **Home > Modify > Copy** drop-down > **Rotate** on the ribbon or enter **RO** in the command line.
- Select the circles as shown below, and then right-click to accept.

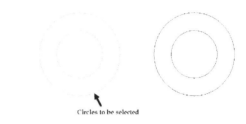

Circles to be selected

- Select the center of the other circle as the base point.

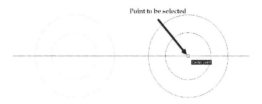

Point to be selected

- Type -90 as the rotation angle and press ENTER; the selected objects are rotated by 90 degrees.

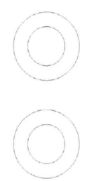

The Scale tool

The **Scale** tool changes the size of objects. It reduces or enlarges the size without changing the shape of an object.

- Create a copy of the circles, as shown.

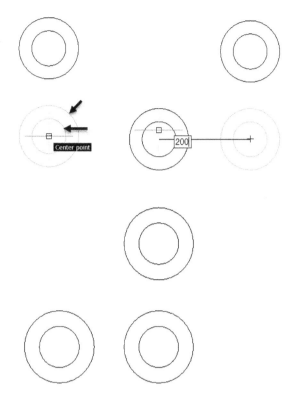

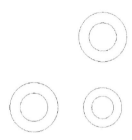

- Likewise, scale the circles located at the top to 0.7.

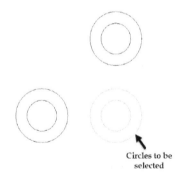

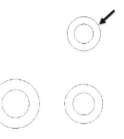

- Click **Home > Modify > Copy** drop-down > **Scale** on the ribbon or enter **SC** in the command line.

- Select the circles as shown below and right-click to accept the selection.

- Click **Home > Draw > Circle > Tangent, Tangent, Radius** on the ribbon.

- Select the two circles shown below to define the tangent points.

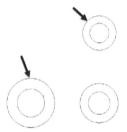

Circles to be
selected

- Select the center point of the selected circles as the base point.

- Type 0.8 as the scale factor and press ENTER.

- Type 150 as the radius of the circle and press ENTER.

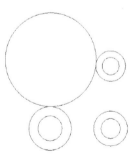

- Likewise, create other circles of radius 100 and 120.

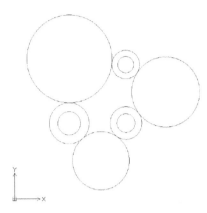

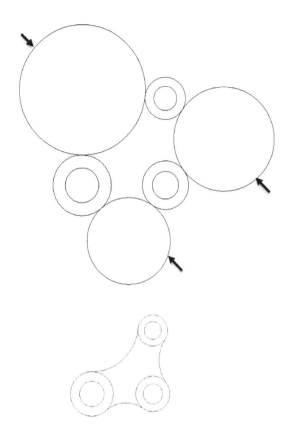

The Trim tool

✁ When an object intersects with another object, you can remove its unwanted portion by using the **Trim** tool. To trim an object, you must first activate the **Trim** tool, and then select the cutting edge (intersecting object) and the portion to be removed. If there are multiple intersection points in a drawing, you can press ENTER; all the objects in the drawing objects act as 'cutting edges.'

- Click **Home > Modify > Trim** drop-down > **Trim** on the ribbon or enter **TR** in the command line.

  Now, you must select the cutting edges.
- Press ENTER to select all the objects as the cutting edges.

  Now, you must select the segments to be removed.
- Select the large circles one by one; the circles are trimmed.

The Power Trim Tool

⊩ The **Power Trim** tool helps you to trim the entities of the drawing by pressing the left mouse button and dragging.

- On the ribbon, click **Home > Modify > Power Trim**.
- Place the mouse button in the center of the drawing.
- Press and hold the left mouse button and drag the pointer across the two circles, as shown.

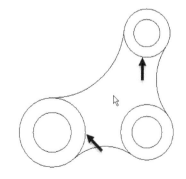

- Release the left mouse button.
- Right click and select Enter.

- Save and close the drawing.

The Extend tool

The **Extend** tool is similar to the **Trim** tool, but its use is the opposite of it. This tool is used to extend lines, arcs, and other open entities to connect to other objects. To do so, you must select the boundary up to which you want to extend the objects, and then select the objects to be extended.

- Start a new drawing.
- Create a sketch, as shown below, using the **Line** tool.

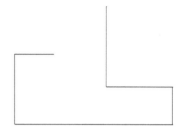

- Click **Home > Modify > Trim > Extend** on the ribbon or enter **EX** in the command line.
- Select the vertical line as the boundary edge. Next, right-click.
- Select the horizontal open line. The line is extended up to the boundary edge.

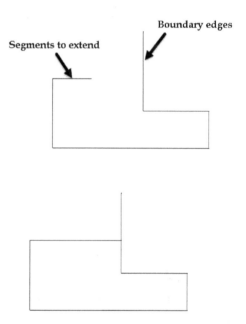

The Corner Trim tool

The **Corner Trim** tool trims and extends elements to form a corner.

- Create a circle and lines, as shown in the figure.
- On the ribbon, click **Home > Modify > Trim** drop-down > **Corner Trim**.

- Click on the circle and horizontal line at the locations, as shown in the figure.

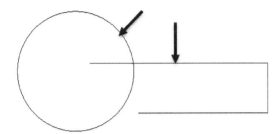

The circle and the line are trimmed to form a corner.

- Click on the trimmed circle and the lower horizontal line.

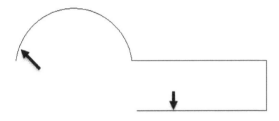

The circle and the line are extended to form a corner.

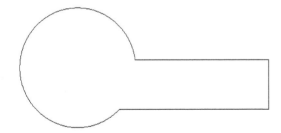

The Fillet tool

The **Fillet** tool converts the sharp corners into round corners. You must define the radius and select the objects forming a corner. The following figure shows some examples of rounding the corners.

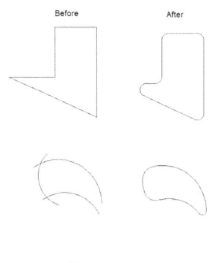

Before   After

- Start a new drawing.
- Type **Limmax** or **GETXYURBNDS** in the command line and press ENTER.
- Set the maximum limit to 100,100 and press ENTER.
- Click **View > Navigate > Zoom Window > Zoom Fit** on the ribbon.
- Click **Home > Draw > Polyline** on the ribbon.
- Define the start point as 20, 50.
- Draw the lines, as shown below.

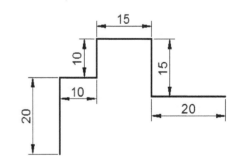

- Right-click and select **Enter**.
- Click **Home > Modify > Trim** drop-down > **Fillet** on the ribbon or enter **F** in the command line.
- Type **R** or **Radius** in the command line.
- Type **5** and press ENTER.

- Select the vertical and horizontal lines, as shown below.

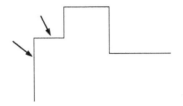

Notice that a fillet is created.

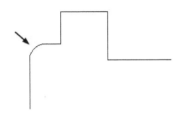

The Chamfer tool

The **Chamfer** tool replaces the sharp corners with an angled line. This tool is similar to the **Fillet** tool, except that an angled line is placed at the corners instead of rounds.

- Click **Home** > **Modify** > **Trim** drop-down > **Chamfer** on the ribbon or enter **CHA** in the command line.

- Follow the prompt sequence given next:

  Options: Angle, Distance, mEthod, Multiple, Polyline, Trim mode, Undo or Select first line >>
  Type in **D** or **Distance** in the command line.

  Default: 0
  Specify first distance >> Enter **8** as the first chamfer distance and press ENTER.
  Default: 8

Specify second distance >> Press ENTER to accept 8 as the second chamfer distance.

Options: Angle, Distance, mEthod, Multiple, Polyline, Trim mode, Undo or Specify first line >> Select the vertical line on the right-side.

Shift + select to apply corner **or**
Specify second line >> Select the horizontal line connected to the vertical line.

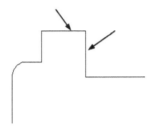

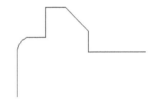

The Mirror tool

The **Mirror** tool creates a mirror image of objects. You can create symmetrical drawings using this tool. Activate this tool and select the objects to mirror, and then define the 'mirror line' about which the objects are mirrored. You can define the mirror line by either creating a line or selecting an existing line.

- Click **Home** > **Modify** > **Copy** drop-down > **Mirror** on the ribbon or enter **MI** in the command line.
- Select the drawing by clicking on it, and then press Enter.

- Select the first point of the mirror line, as shown below.

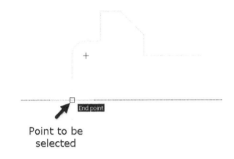

Point to be selected

- Make sure that the **Ortho** icon on the status bar is active.

- Move the pointer toward the right and click.

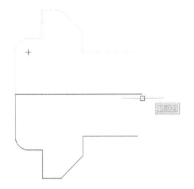

- Type **N** or **No** in the command line to retain the source objects.

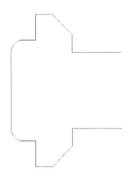

- Click **Home > Draw > Arc > Start, End, Direction** on the ribbon.

- Select the start point of the arc, as shown.

- Select the endpoint of the arc, as shown.

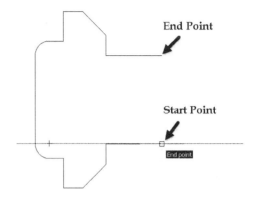

End Point

Start Point

End point

- Make sure that the **Ortho** icon is active.

- Move the pointer toward the right and click.

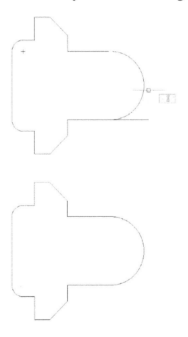

The Explode tool

The **Explode** tool explodes a group of objects into individual objects. For example, when you create a drawing using the **Polyline** tool, it acts as a single object. You can explode a polyline or rectangle or any group of objects using the **Explode** tool.

- Click on the portion of the drawing created using the **Polyline** tool; notice that the complete polyline is selected as a single object.

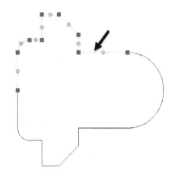

- Click **Home > Modify > Explode** on the ribbon or enter **X** in the command line.
- Select the polylines from the drawing.

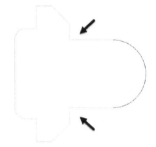

The polyline is exploded into individual objects.

Now, you can select the individual objects of the polyline.

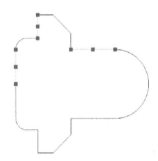

The Stretch tool

The **Stretch** tool lengthens or shortens drawings or parts of drawings. Note that you cannot stretch circles using this tool. In addition to that0, you must select the

portion of the drawing to be stretched by dragging a window.

- Click **Home > Modify > Stretch** on the ribbon or enter **STRETCH** in the command line.
- Create a crossing window to select the objects of the drawing.

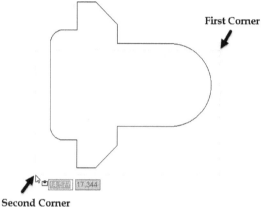

First Corner

Second Corner

- Press ENTER (or) right-click to accept the selection.
- Select the from point, as shown below.

From Point

- Move the pointer downward and click to stretch the drawing.

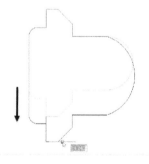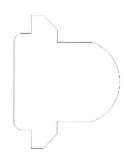

- Save and close the file.

Creating Circular Patterns

The **Pattern** tool can be used to create an arrangement of objects around a point in a circular form. The following example shows you to create a polar array.

- Create two concentric circles of 140 and 50 diameters.

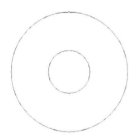

- Type **C** in the command line and press ENTER.
- Press and hold the Shift key, right-click, and select **Quadrant** from the shortcut menu.
- Select the quadrant point of the circle, as shown below.

- Type 30 as radius and press ENTER.

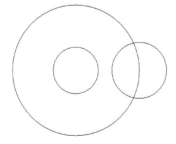

- Click **Home > Modify > Trim** drop-down **> Trim** on the ribbon.

- Select the large circle as the cutting edge and right-click.
- Select the circle on the quadrant as the segment to remove.

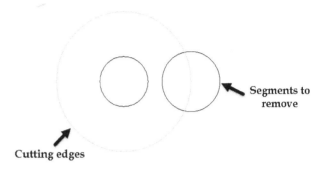

- Press Enter.
- Click **Home > Modify > Pattern** on the ribbon or Type **PAT** or **PATTERN** in the command line; the **Pattern** dialog appears on the screen.
- Click the **Circular** tab and click on the **Specify entities** button under the **Selection** section.

- Select the arc created after trimming the circle. Next, right-click to accept the selection.

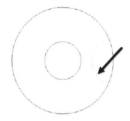

- Click the **Pick Center Point** button under the **Axis point** section.

- Select the center of the large circle as the axis point of the pattern.

- Select **Fill Angle and Angle Between Elements** from the **Base pattern** on the drop-down under the **Settings** section.

- Set the **Angle between** and **Fill angle** values to 90 and 360.

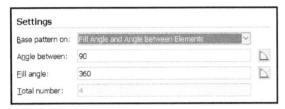

- Click the **Preview** button to preview the pattern. Next, press Esc.

- Click **OK** to create the circular pattern.

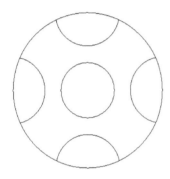

- Click **Home > Modify > Power Trim > Trim** on the ribbon.

- Press ENTER to select all objects as cutting edges.

- Trim the unwanted portions, as shown below.

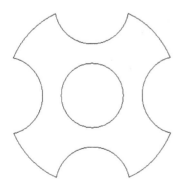

The Offset tool

The **Offset** tool creates parallel copies of elements such as lines, polylines, circles, and arc. To create a parallel copy of an object, first, you must define the offset distance and then select the object. Next, you must define the side in which the parallel copy will be placed.

- Create the drawing shown below using the **Polyline** tool. Do not add dimensions.

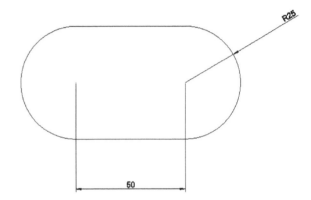

- Click **Home > Modify > Offset** on the ribbon or enter **O** in the command line.

- Type **20** as the offset distance and press ENTER.

- Select the polyline loop.

- Click outside the loop to create a parallel copy.

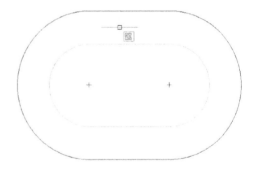

- Type **Exit** in the command line.

- Click **Home > Layers > Layer Manager** on the ribbon (or) type **LA** in the command line; the **Layer Manager** appears.

- Click the **New** button on the **Layer Manager** — Enter **Centerline** in the **Name** field.

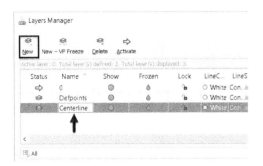

- Click the **Activate** ⇨ icon.
- Click the **LineStyle** field of the current layer; a drop-down appears.
- Select the **Other** option; the **Line Style** dialog appears.

- On the **Line Style** dialog, click the **Load** button on the right; the **Load LineStyles** dialog appears.

- Select the **CENTER** LineStyle from this dialog.

- Click **OK**. The linestyle is added to the **Line Style** dialog.
- Select the **CENTER** linestyle from the **Line Style** dialog.
- Type 0.5 in the **Global Scale** and **Entities Scale** boxes, and then click **OK**.

- Click **OK** on the **Layer Manager**.
- Click the **Offset** button on the **Modify** panel.
- Type **L** or **Layer** in the command line.
- Type **A** or **Active** in the command line; this ensures that the offset entity is created with the currently active layer properties. If you type **Source**, the offset entity is created with the properties of the source object.
- Type **10** as the offset distance and press ENTER.
- Select the outer loop of the drawing.
- Move the pointer inwards and click to create the offset entity.

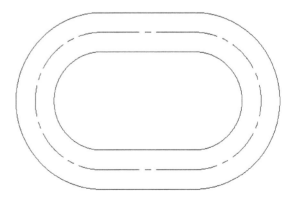

- Click on the **Layer** drop-down on the **Layer** panel of the ribbon.
- Select the **0** layer from the drop-down.

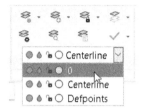

- Create a circle of 12 mm in diameter.

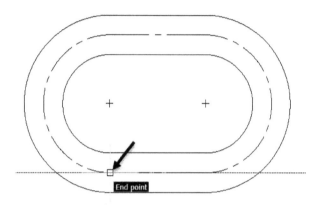

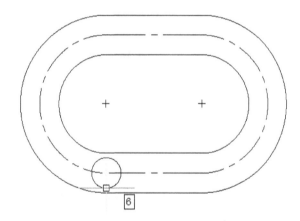

Creating Path Patterns

The **Pattern** tool can create an array of objects along a path (line, polyline, circle, helix, and spline).

- Click **Home** > **Modify** > **Pattern** on the ribbon or enter **PAT** in the command line.
- Click the **Path** tab on the **Pattern** dialog.
- Click the **Specify entities** button, select the circle, and press ENTER.
- Click the **Specify Path** button and select the centerline as the path.

- Click the **Base pattern on** drop-down and select the **Divide Equally** option. Now you must enter the number of elements in the path pattern. If you select the **Measure Equally** method, you must enter the distance between the elements in the path pattern.

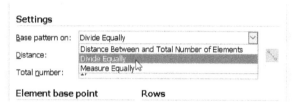

- Set the **Total number** to 12.
- Uncheck **Use last entity selected** under the **Element base point** section.
- Click the **Specify base point** button and click on the center point of the circle.
- Click the **Preview** button to preview the path pattern.

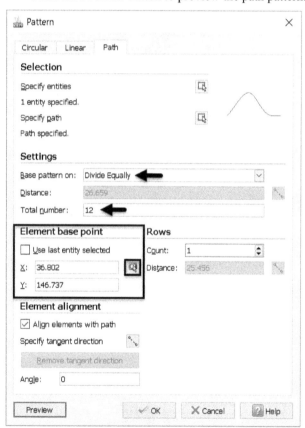

- Press **Esc** to return to the dialog.

Notice that the **Align elements with path** option is checked by default. As a result, the elements are aligned with the path. If you deactivate this button, the elements are not aligned with the path.

Align elements with   Align elements with
path selected   path deselected

- Click **OK** on the **Pattern** dialog to create the path pattern.

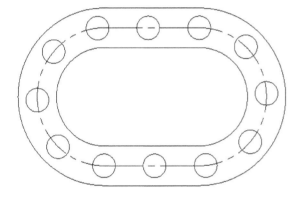

- Save and close the file.

Creating Linear Patterns

The **Pattern** tool allows you to create an array of objects along the X and Y directions.

- Open a new DraftSight file and draw the sketch shown below. Do not add dimensions. (refer to the **Drawing Rectangles** and **Drawing Circles** section in Chapter 2 to know the procedure to draw the rectangle and circle)

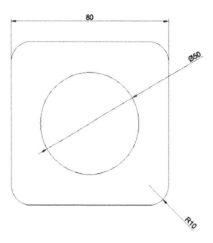

- Activate the **Circle** command and select the center point of the lower-left fillet.
- Type 5 and press ENTER.

- Click **Home > Modify > Pattern** on the ribbon or enter **PAT** or **PATTERN** in the command line.
- On the **Pattern** dialog, click the **Linear** tab.
- Click the **Specify Entities** button.
- Select the small circle and press ENTER.
- Set the **Vertical axis** to 2 under **Settings > Number of elements on** section.
- Set the **Horizontal axis** to 2.
- Set the values in the **Spacing between elements on** section as given below.
  - Set the **Vertical axis** to 60.
  - Set the **Horizontal axis** to 60.

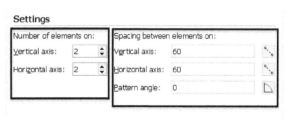

- Click **OK** on the dialog.

Editing Using Grips

When you select objects from the graphics window, small squares appear on them. These squares are called grips. You can use these grips to stretch, move, rotate, scale and mirror objects change properties, and perform other

editing operations. Grips displayed on selecting different objects are shown in the figure.

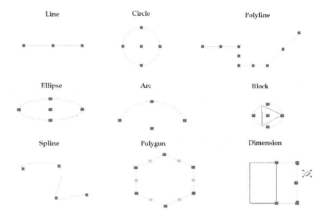

The following table gives you the details of the editing operations that can be performed when you select and drag grips.

Object	Grip	Editing Operation
Circle	Grip on circumference	**Scale**: Select anyone of the grips on the circumference and move the pointer; the size of the circle is modified.
	Center point grip	**Move**: Select the center grip of the circle and move the pointer.

Arc	Grip on circumference	**Stretch**: Select the grip on the circumference and move the pointer.
	Center point grip	**Move**: Select the center grip of the arc and move the pointer.
Line	Midpoint Grip	**Move**: Select the Midpoint grip and move the pointer
	Endpoint Grip	**Stretch/Lengthen**: Select an endpoint grip and move the pointer.

Polylines, Rectangles, Polygons	Corner Grips	**Stretch**: Select the corner grips and move the pointer. **Add/Remove Vertex**: Place the pointer on the corner grip and select Add Vertex/Remove Vertex.
Ellipse	Center Grip	**Move**: Select the center grip and move the pointer.
	Grips on circumference	**Stretch**: Select a grip on the circumference and move the pointer.

Spline	Fit Points	**Stretch**: Select a grip on the spline and move the pointer. 

Revision Clouds

Revision clouds are used to highlight the areas in a drawing. You can create revision clouds using three different tools.

Example 1:

- Start a new drawing using the standardiso template.
- On the ribbon, click **Annotate > Markup > Cloud > Rectangular** .
- Click an arbitrary point to specify the first corner.
- Move the pointer upward or downward and click to specify an opposite corner.

- Select the cloud and notice the grip. You can use it to move the cloud.

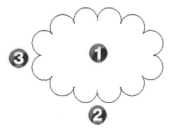

Example 2:

- On the ribbon, click **Annotate > Markup > Cloud > Elliptical** .
- Click an arbitrary point to specify the start axis point.
- Move the pointer downwards and click to specify the end axis point.
- Likewise, move the pointer towards the right and click to specify the third axis point.

Example 3:

- On the ribbon, click **Annotate > Markup > Revision**

  Cloud > Freehand  .

- Specify the start point of the cloud.

- Move the pointer and draw a sketch of your own.

- Move the pointer onto the start point to close the cloud.

- Type **C** or **Close** in the command line. Next, press ENTER.

Exercises

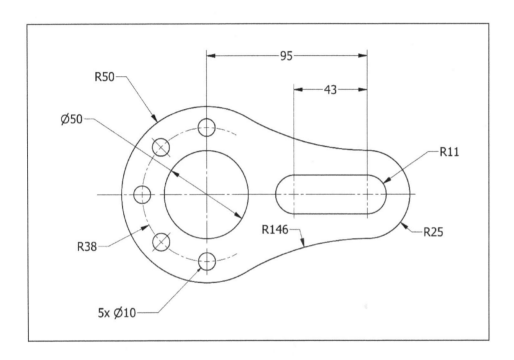

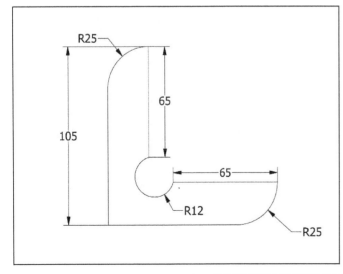

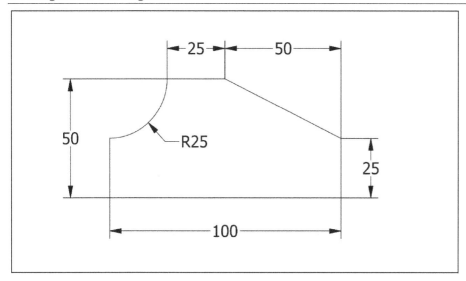

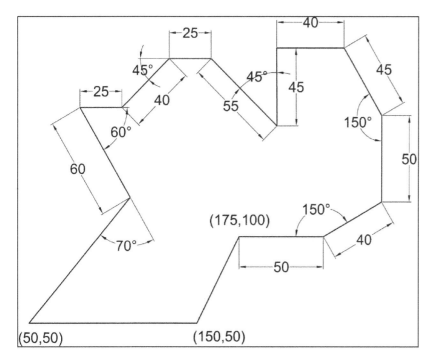

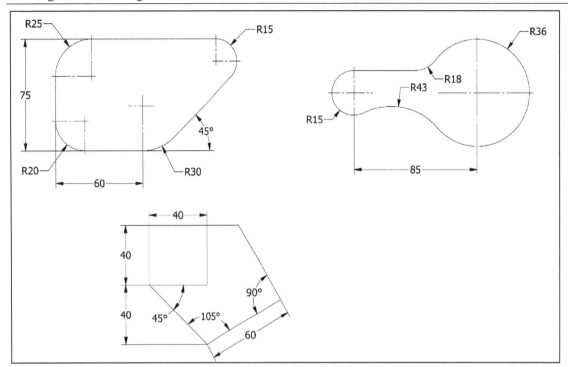

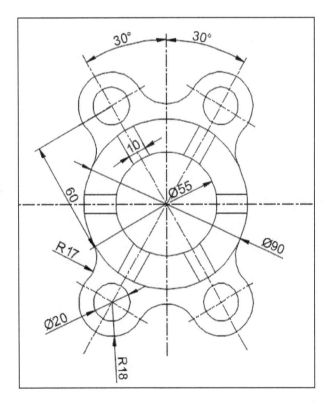

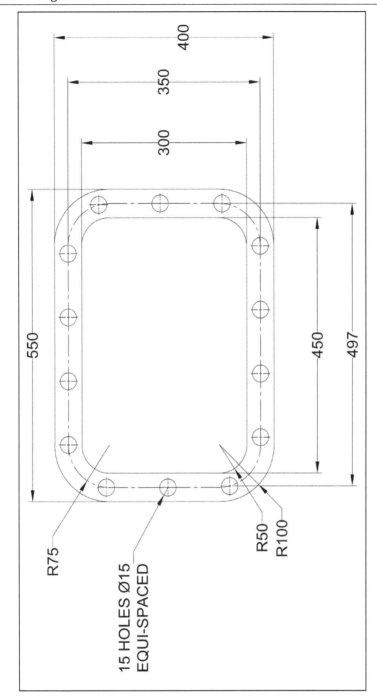

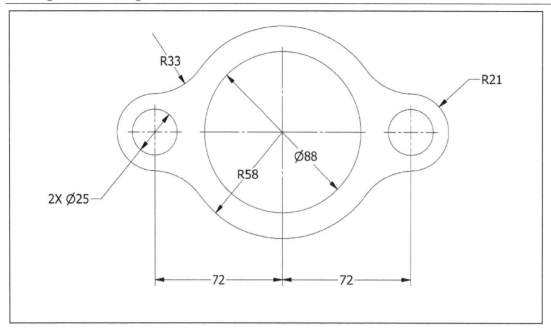

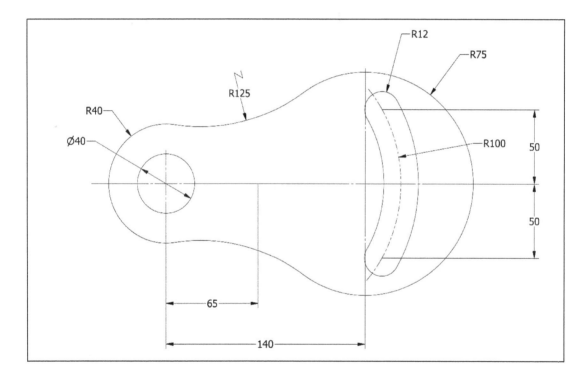

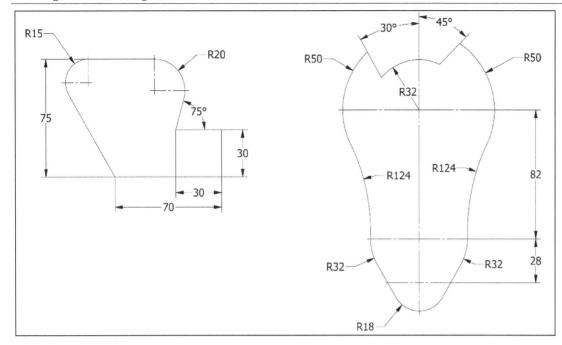

Chapter 5: Multi View Drawings

In this chapter, you learn to create:

- **Orthographic Views**
- **Auxiliary Views**
- **Named Views**

Multi-view Drawings

To manufacture a component, you must create its engineering drawing. The engineering drawing consists of various views of the object, showing its true shape and size so that it can be clearly dimensioned. This can be achieved by creating the orthographic views of the object. In the first section of this chapter, you learn to create orthographic views of an object. The second section introduces you to auxiliary views. The auxiliary views clearly describe the features of a component, which are located on an inclined plane or surface.

Creating Orthographic Views

Orthographic Views are standard representations of an object on a sheet. These views are created by projecting an object onto three different planes (top, front, and side planes). You can project an object by using two different methods: **First Angle Projection** and **Third Angle Projection**. The following figure shows the orthographic views that are created when an object is projected using the **First Angle Projection** method.

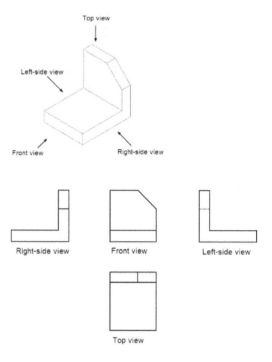

The following figure shows the orthographic views that are created when an object is projected using the **Third Angle Projection** method.

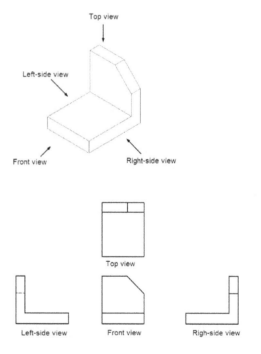

Example:

In this example, you create the orthographic views of the part shown below. The views are created by using the **Third Angle Projection** method.

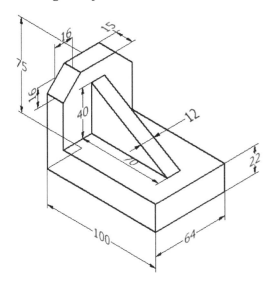

- Open a new drawing using the **Standardiso.dwt** template.

- Click the **Layer Manager** button on the **Layers** panel; the **Layer Manager** appears.

- Click the **New** button on the **Layer Manager** to create new layers.

- Create two new layers with the following properties.

Layer Name	Lineweight	Linetype
Construction	0.00 mm	Continuous
Object	0.30 mm	Continuous

- Right-click on the **Construction** layer and select **Activate Layer**.

- Close the **Layer Manager**.

- Activate the **Ortho** icon on the status bar.

- Click **Navigate > Zoom > Zoom Fit** on the **View** tab.

Next, you need to draw construction lines. They are used as references to create actual drawings. You create these construction lines on the **Construction** layer so that you can hide them when required.

- Click **Home > Draw > Infinite line** on the ribbon or enter **XLINE** in the command line.

- Click anywhere in the lower-left corner of the graphics window.

- Move the pointer upward and click to create a vertical construction line.

- Move the pointer toward the right and click to create a horizontal construction line.

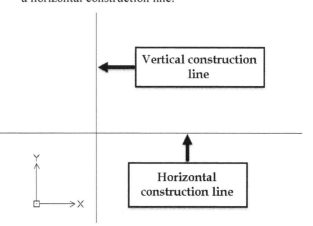

- Press ENTER to exit the tool.

- Click the **Offset** button on the **Modify** panel.

- Type 100 as the offset distance and press ENTER.

- Select the vertical construction line.

- Move the pointer toward the right and click to create an offset line.

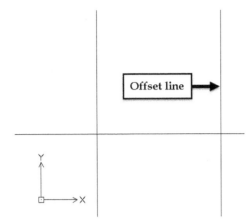

- Right-click and select **Enter** to exit the **Offset** tool.
- Press the SPACEBAR on the keyboard to start the **Offset** tool again.
- Type 75 as the offset distance and press ENTER.
- Select the horizontal construction line.
- Move the pointer above and click to create the offset line.

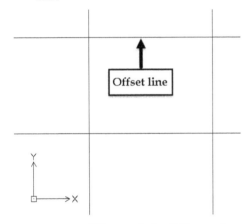

- Press ENTER to exit the **Offset** tool.
- Likewise, create other offset lines, as shown below. The offset dimensions are displayed in the image. Do not add dimensions to the lines.

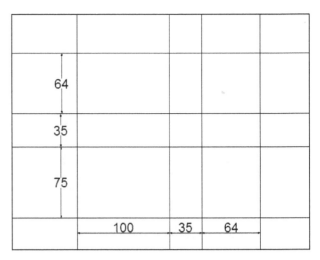

- Activate the **Object** layer.

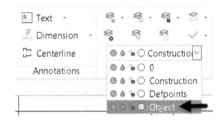

  Now, you must create the object lines.

- Click the **Lineweight** ☰ icon on the **Properties** group of the **Home** ribbon tab.
- Click **Drafting styles** option on the left of the **Options** dialog.
- Expand **Active Drafting styles > Line Font > Line Weight: By Layer** on the dialog.
- Check the **Display weight in graphics area** option.

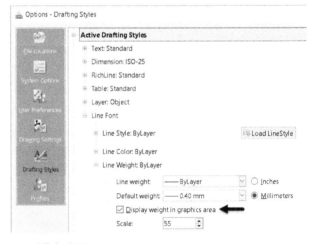

- Click **OK**.
- Click the **Line** button on the **Draw** panel.

- Create an outline of the front view by selecting the intersection points between the construction lines.

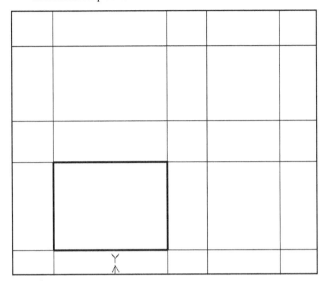

- Right-click and select **Enter** to exit the **Line** tool.
- Likewise, create the outlines of the top and side views.

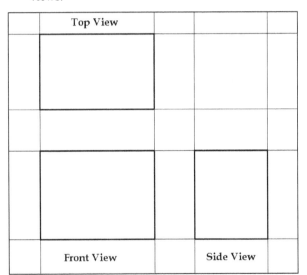

Next, you must turn off the **Construction** layer.

- Click on the **Layers Manager** drop-down in the **Layers** panel.
- Click the green dot of the **Construction** layer; the layer is turned off.

- Use the **Offset** tool and create two parallel lines on the front view, as shown below.

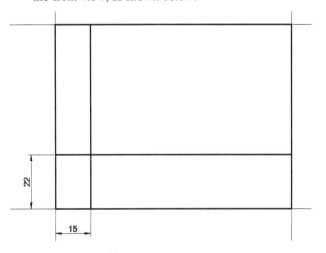

- Use the **Trim** tool and trim the unwanted lines of the front view, as shown below.

- Use the **Offset** tool to create the parallel line, as shown below.

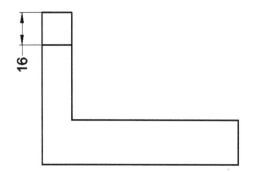

- Use the **Offset** tool and create offset lines in the Top view, as shown below.

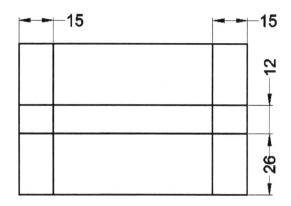

- Use the **Trim** tool and trim the unwanted objects.

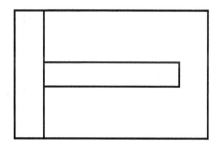

- Create other offset lines and trim the unwanted portions, as shown below.

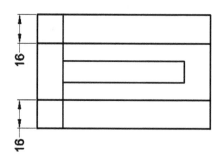

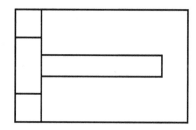

- Deactivate the **Ortho** icon on the status bar.
- Click on the **Layers Manager** drop-down in the **Layers panel**.
- Click the green dot of the **Construction** layer; the layer is turned on.

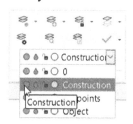

- Click the **Offset** button on the **Modify** panel.
- Type **T** or **Through point** in the command line and press ENTER.
- Select the vertical construction line, as shown.

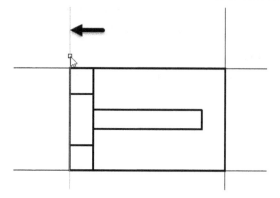

- Next, select the endpoint on the Top View, as shown; an offset of the construction line is created.

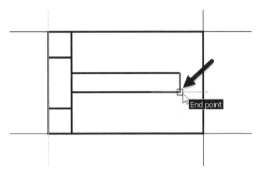

- Press the SPACEBAR on the keyboard to start the **Offset** tool again.
- Type 62 as the offset distance and press ENTER.
- Select the horizontal construction line, as shown.

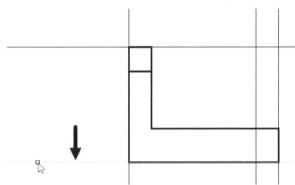

- Move the pointer above and click to create the offset line.

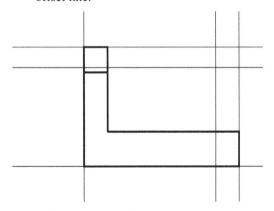

- Right-click and press Enter.
- Turn on the object layer, by clicking the **Layers Manager** drop-down > **Object** on the **Layer** panel.
- Click the **Line** button on the **Draw** panel.

- Place the pointer on the intersection of the newly created vertical construction line, as shown.

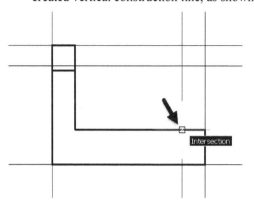

- Move the pointer diagonally upward and click on the intersection of the vertical line and horizontal construction line, as shown.

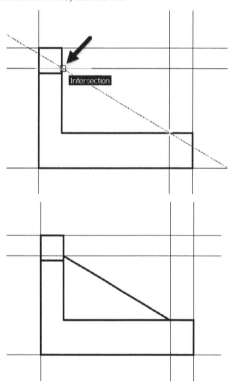

Next, you must create the right side view. To do this, you must draw a 45- degree miter line and project the measurements of the top view onto the side view.

- Click on the **Layers Manager** drop-down in the **Layers** panel.

- Select the **Construction** layer from the **Layers Manager** drop-down to set it as the current layer.
- Draw an inclined line by connecting the intersection points of the construction lines, as shown below.

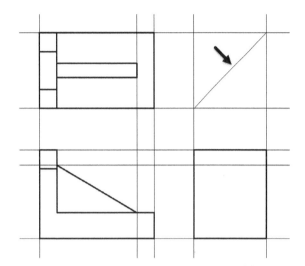

- Click the **Line** drop-down > **Infinite Line** button on the **Draw** panel.
- Type H or Horizontal in the command line and press Enter.
- Create the projection lines, as shown below.

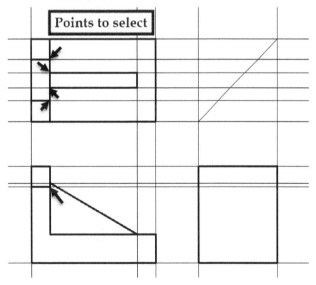

- Right-click to exit the **Infinite Line** tool.
- Press ENTER and type V or Vertical in the command line. Then press Enter.
- Create vertical projection lines, as shown below.

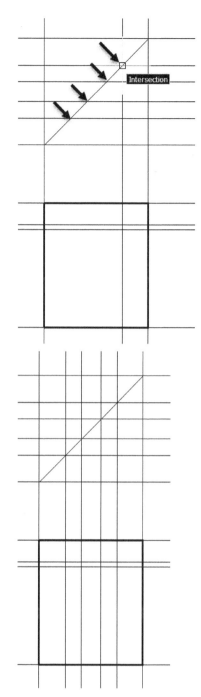

- Use the **Trim** tool trim the extending portions of the construction lines.

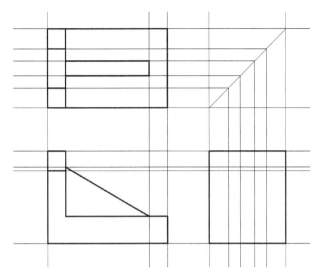

- Set the **Object** layer as current.
- Click the **Offset** button on the **Modify** panel.
- Type **T** or **Through point** in the command line and press Enter.
- Select the lower horizontal line of the side view.

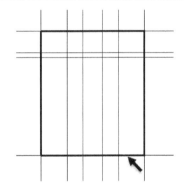

- Select the endpoint on the front view, as shown below.

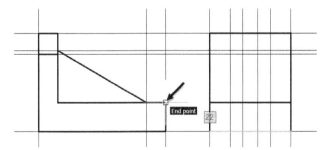

- Type **E** or **Exit** in the command line. Press ENTER.
- Use the **Line** tool and create the objects in the side view, as shown below.

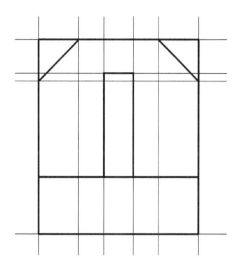

- Turn off the **Construction** layer by clicking on the green dot of the **Construction** layer.

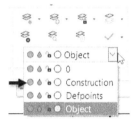

- Trim the unwanted portions on the right side view.

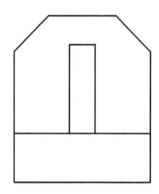

The following figure shows the drawing after creating all the views.

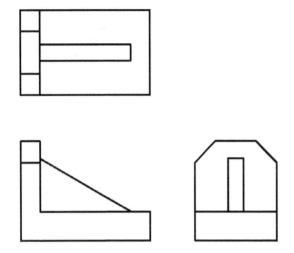

- Save the file as **ortho_views.dwg**. Close the file.

Creating Auxiliary Views

Most of the components are represented by using orthographic views (front, top, and/or side views). But many components have features located on inclined faces. You cannot get the true shape and size for these features by using the orthographic views. To see an accurate size and shape of the inclined features, you must create an auxiliary view. An auxiliary view is created by projecting the component onto a plane other than horizontal, front, or side planes. The following figure shows a component with an inclined face. When you create orthographic views of the component, you are not able to get the true shape of the hole on the inclined face.

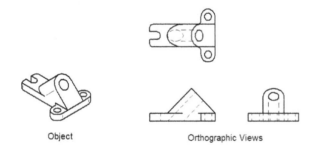

Object   Orthographic Views

To get the actual shape of the hole, you must create an auxiliary view of the object, as shown below.

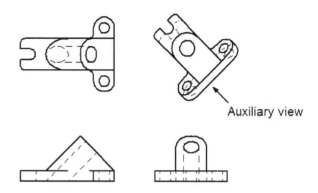

Auxiliary view

Example:

In this example, you create an auxiliary view of the object shown below.

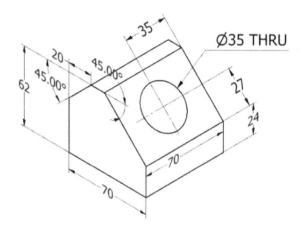

Ø35 THRU

- Open a new DraftSight file.
- Create four new layers with the following properties.

Layer Name	Lineweight	Linetype
Construction	0.00 mm	Continuous
Object	0.50 mm	Continuous
Hidden	0.30 mm	HIDDEN
Centerline	0.30 mm	CENTER

- Select the **Construction** layer from the **Layers Manager** drop-down in the **Layers** panel.
- Activate the **Ortho** ⊡ icon on the status bar.
- Click **Home > Draw > Infinite line** on the ribbon or enter **XLINE** in the command line.

- Click anywhere in the lower-left corner of the graphics window.
- Move the pointer upward and click to create a vertical construction line.
- Move the pointer toward the right and click to create a horizontal construction line.

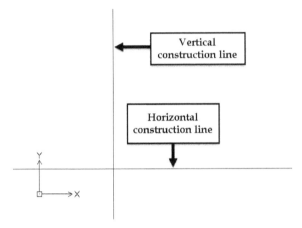

- Press ENTER to exit the tool.
- Click the **Offset** button on the **Modify** panel.
- Type 122 as the offset distance and press ENTER.
- Select the horizontal construction line.
- Move the pointer upward and click to create an offset line.

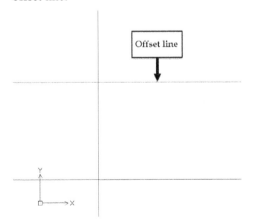

- Right-click and select **Enter** to exit the **Offset** tool.
- Create a rectangle at the lower-left corner of the graphics window, as shown in the figure.

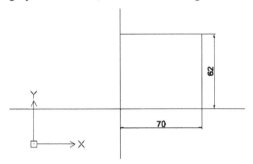

- Select the rectangle and click the **Copy** button on the **Modify** panel.
- Select the lower-left corner of the rectangle as the from point.
- Make sure that the **Ortho** icon is activated.
- Move the pointer upward and type **25** in the command line — next, press ENTER.
- Press ESC to exit the **Copy** tool.

- Click the **Copy** drop-down > **Rotate** button on the **Modify** panel and select the copied rectangle. Press ENTER to accept.
- Select the lower right corner of the copied rectangle as the pivot point.
- Type 45 as the angle and press ENTER.

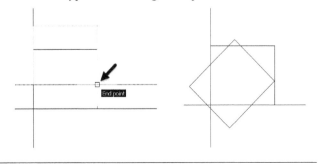

- Press Esc to exit the **Rotate** tool.
- Click the **ESnap** 🔲 and **ETrack** 🔳 icons on the Status Bar.
- Activate the **Rectangle** command.
- Place the pointer on the intersection of the horizontal and vertical lines, as shown.

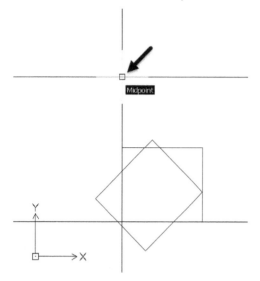

- Type **D** or **Dimension** in the command line and press Enter.
- Type 70 in the command line as the horizontal dimension and press Enter.
- Again, type 70 in the command line as the vertical dimension and press Enter.

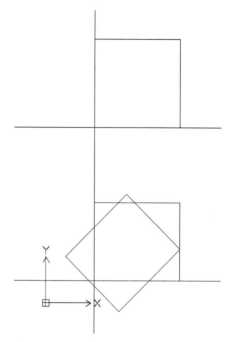

The rectangle located at the top is considered as top view and the below one as the front view.

- Click the **Explode** button on the **Modify** panel and select the newly created rectangle. Next, right-click to explode the rectangle.
- Activate the **Offset** tool and type **T** or **Through point** in the command line — next, press ENTER.
- Select the left vertical line of the top rectangle.
- Select any one of the through points, as shown; the selected vertical line is offset through the selected point.
- Again, select the left vertical line.
- Move the pointer, and then select the remaining through point.

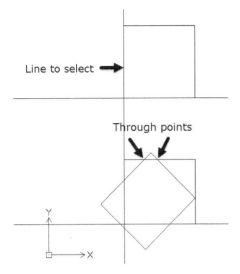

- Select the **Object** layer from the **Layers Manager** drop-down in the **Layers** panel.
- Click the **Lineweight** ≡ icon on the **Properties** group of the **Home** ribbon tab.
- Click **Drafting styles** option on the left of the **Options** dialog.
- Expand **Active Drafting styles > Line Font > Line Weight: By Layer** on the dialog.
- Check the **Display weight in graphics area** option.

- Click **OK**.
- Activate the **Line** tool and select the intersection points on the front view, as shown.

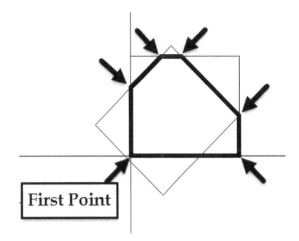

- Likewise, create the object lines in the top view, as shown below.

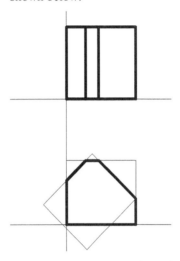

- Select the **Construction** layer from the **Layers** panel.
- Click the **Infinite Line** button on the **Draw** panel.
- Type **O** or **Offset** in the command line. Next, press ENTER.
- Type **P** in the command line. Next, press ENTER.
- Select the inclined line on the front view. Next, select the intersection point, as shown below.

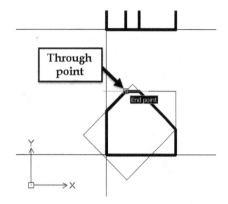

- Likewise, create other construction lines, as shown below.

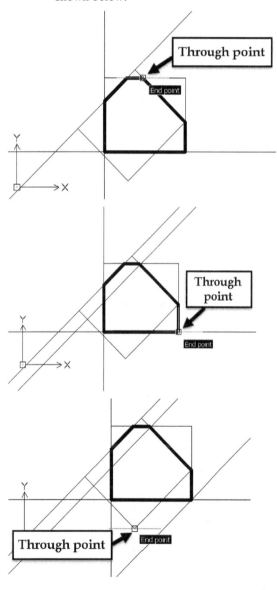

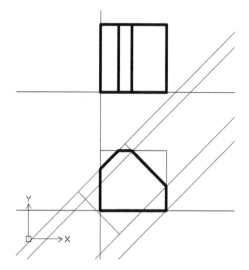

- Press Esc to deactivate the **Infinite Line** command.

- Activate the **Infinite Line** command and type **O** or **Offset** in the command line — next, press ENTER.

- Type 80 and press ENTER.

- Select the inclined line on the front view, as shown.

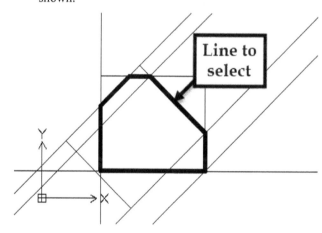

- Move the pointer toward the right and click to create the construction line.

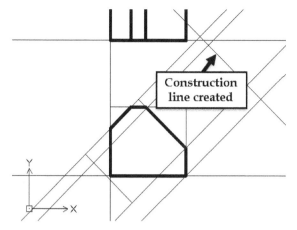

- Create other construction lines, as shown. The offset dimensions are given in the figure.

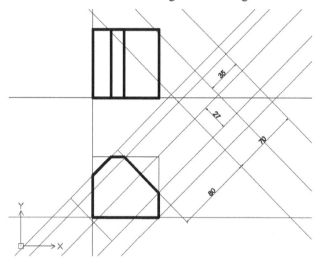

- Set the **Object** layer as the current layer. Next, create the object lines using the intersection points between the construction lines.
- Use the **Circle** tool and create a circle of 35 mm in diameter.

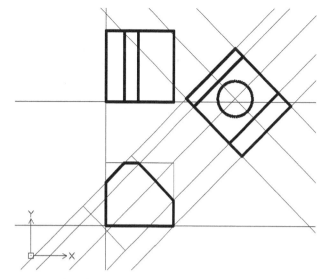

- Set the **Construction** layer as the current layer.
- Create projection lines from the circle.

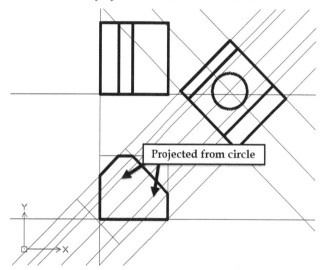

- Set the **Hidden** layer as a current layer,
- Create the hidden lines, as shown.

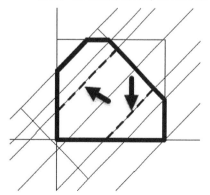

- Set the **Centerline** layer as a current layer,
- Create the centerlines, as shown.

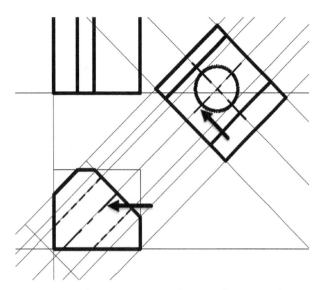

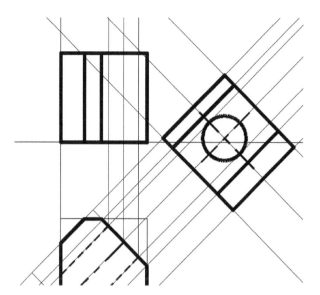

- Set the **Construction** layer as the current layer.
- On the ribbon, click **Home** > **Draw** > **Line** drop-down > **Ray** .
- Select the intersection point of the hidden line and object line, as shown.
- Move the pointer upward and click to create a ray.

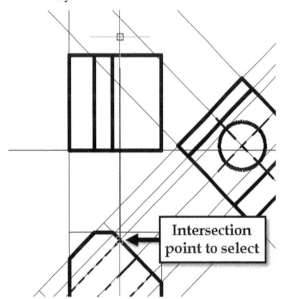

Intersection point to select

- Press ENTER twice.
- Likewise, create two more rays, as shown.

- Create a horizontal construction line passing through the midpoint of the top view, as shown.

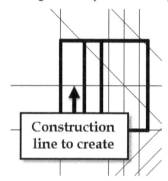

Construction line to create

- Set the **Object** layer as a current layer,
- On the ribbon, click **Home** > **Draw** > **Ellipse** drop-down > **Axis, End** .
- Specify the first and second points, as shown.

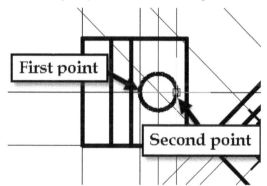

First point

Second point

- Move the point downward, type-in 17.5, and then press ENTER.

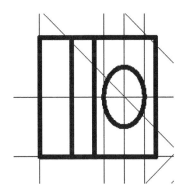

- Set the **Centerline** layer as a current layer,
- Create the remaining centerlines.
- The drawing after hiding the **Construction** layer is shown next.
- Save the file as auxiliary_views.dwg.

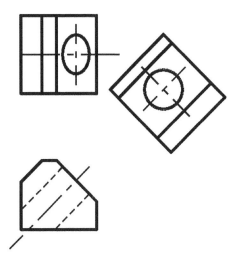

Creating Named views

While working with a drawing, you may need to perform numerous zoom and pan operations to view key portions of a drawing. Instead of doing this, you can save these portions with a name. Then, restore the named view and start working on them.

- Open the **ortho_views.dwg** file (The drawing file created in the Orthographic Views section of this chapter).
- Click the **View** tab on the ribbon.
- To create a named view, click **Views > Named Views** on the ribbon; the **Views** dialog appears.

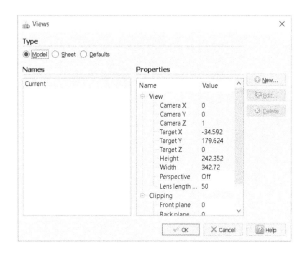

- Click the **New** button on the **Views** dialog; the **View** dialog appears.
- Select the **Specify later** option from the **Boundary** section of the **View** dialog.
- Create a window on the front view, as shown below.

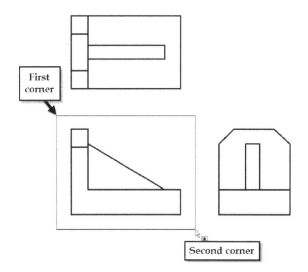

- Press ENTER to accept.
- Enter **Front** in the **Name** box.

- Click **OK** on the **View** dialog.
- Likewise, create the named views for the top and right views of the drawing.

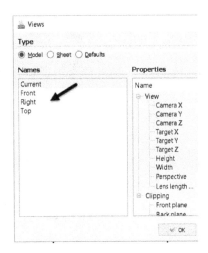

- To set the **Top** view to current, select it from the **Names** tree on the dialog. Next, click **OK** on the **Views** dialog; the **Top** view is zoomed and fitted to the screen.

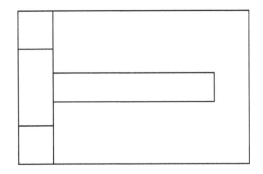

- Save and close the file.

Exercise

Exercise 1

Create the orthographic views of the object shown below.

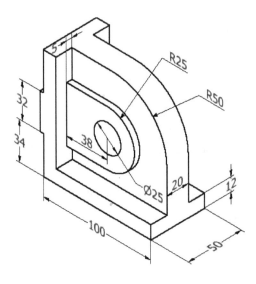

Exercise 2

Create the orthographic views of the object shown below.

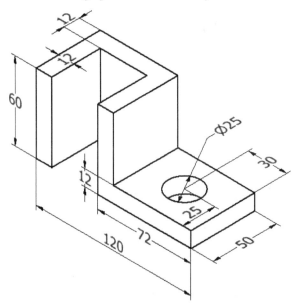

Exercise 3

Create the orthographic and auxiliary views of the object shown below.

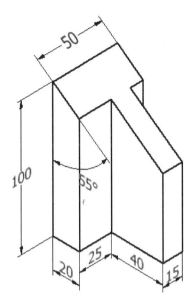

Exercise 4

Create the orthographic and auxiliary views of the object shown below.

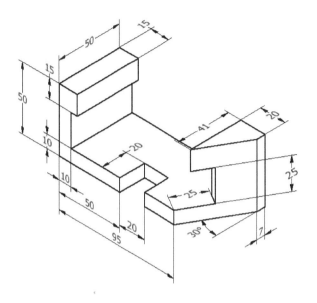

Chapter 6: Dimensions and Annotations

In this chapter, you learn to do the following:

- **Create Dimensions**
- **Create Dimension Style**
- **Add Leaders**
- **Create Centerlines**
- **Add Dimensional Tolerances**
- **Add Geometric Tolerances**
- **Edit Dimensions**

Dimensioning

In previous chapters, you have learned to draw shapes of various objects and create drawings. However, while creating a drawing, you also need to provide the size information. You can provide the size information by adding dimensions to the drawings. In this chapter, you learn how to create various types of dimensions. You also learn about some standard ways and best practices of dimensioning.

Creating Dimensions

In DraftSight, there are many tools available for creating dimensions. You can access these tools from the Ribbon, Command line, and Menu Bar.

The following table gives you the functions of various dimensioning tools.

Tool	Shortcut	Function
Smart	SMARTDIMENSION	This tool creates a dimension based on the selected geometry.

- Create a rectangle, circle, arc, and two intersecting lines, as shown in the previous figure.
- Click **Annotate > Dimensions > Smart** on the ribbon.
- Select a line, move the pointer, and click to create the linear dimension.
- Select a circle, move the pointer, and click to position the diameter dimension.
- Select an arc, move the pointer, and click to position the radial dimension.
- Select an arc, type L, and press Enter. Next, move the pointer and click to position the arc length dimension.

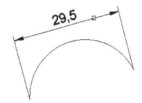

- Select arc, type AN, and press Enter. Next, move the pointer and click to position the angle of the arc.

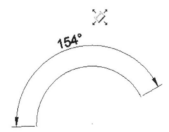

- Select two non-parallel lines and position the angular dimension between them.

Likewise, you can create other types of dimensions using the **Smart** tool.

Linear

	DLI	This tool creates horizontal and vertical dimensions.

- Click **Annotate > Dimensions > Dimension** drop-down> **Linear** on the ribbon.

- Select the first and second points of the dimension.

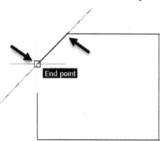

- Move the pointer in a horizontal direction to create a vertical dimension (or) move in the vertical direction to create a horizontal dimension.

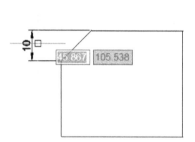 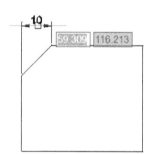

- Click to position the dimension.

Parallel

                DAL   This tool creates a linear dimension parallel to the object.

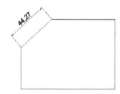

- Click **Annotate > Dimensions > Dimension** drop-down > **Parallel** on the ribbon.

- Select the first and second points of the dimension line (or) press Enter and select the line.

- Move the pointer and click to position the dimension.

Arc Length

                DAR   It dimensions the total or partial length of an arc.

 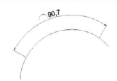

- Click **Annotate > Dimensions > Dimension** drop-down > **Arc Length** on the ribbon.

- Select an arc from the drawing.

- If you want to dimension only a partial length of an arc, type P or Partial, and then press ENTER. Next, select the two points on the arc.

- Move the pointer and click to position the dimension.

Continue

DCO

It creates a linear dimension from the second extension line of the previous dimension.

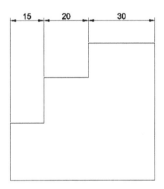

- Create a linear dimension by selecting the first and second points.

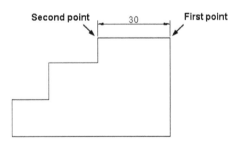

- Click **Annotate > Dimensions > Continue** on the ribbon; a chain dimension is attached to the pointer.

- Select the third and fourth points of the chain dimension.

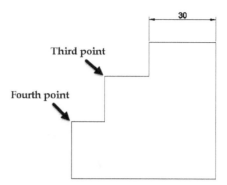

- Position the chain dimension. Next, right-click and select **Enter**.

Baseline

DBA

It creates dimensions by using the previously created dimension, as shown below.

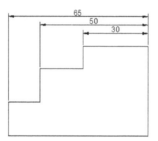

- Create a linear dimension by selecting the first and second points.

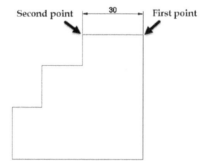

- Click **Annotate > Dimensions > Continue > Baseline** on the ribbon.
- Select the third and fourth points of the baseline dimension. Next, right-click and select **Enter**.

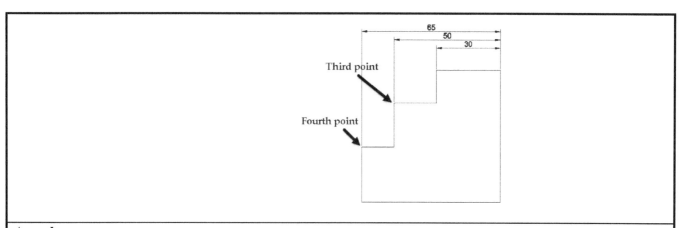

Angular

DAN   It creates an angular dimension.

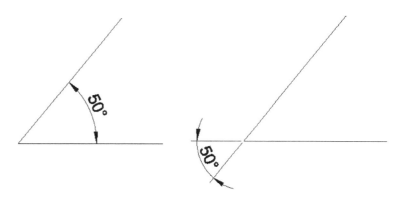

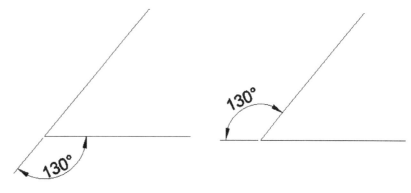

- Click **Annotate > Dimensions > Dimension** drop-down **> Angle** on the ribbon.
- Select the first line and second line.
- Move the pointer and position the angle dimension.
- To create an angle dimension on an arc, select the arc and position the dimension.
- To create an angle dimension on a circle, select two points on the circle and position the angle dimension.

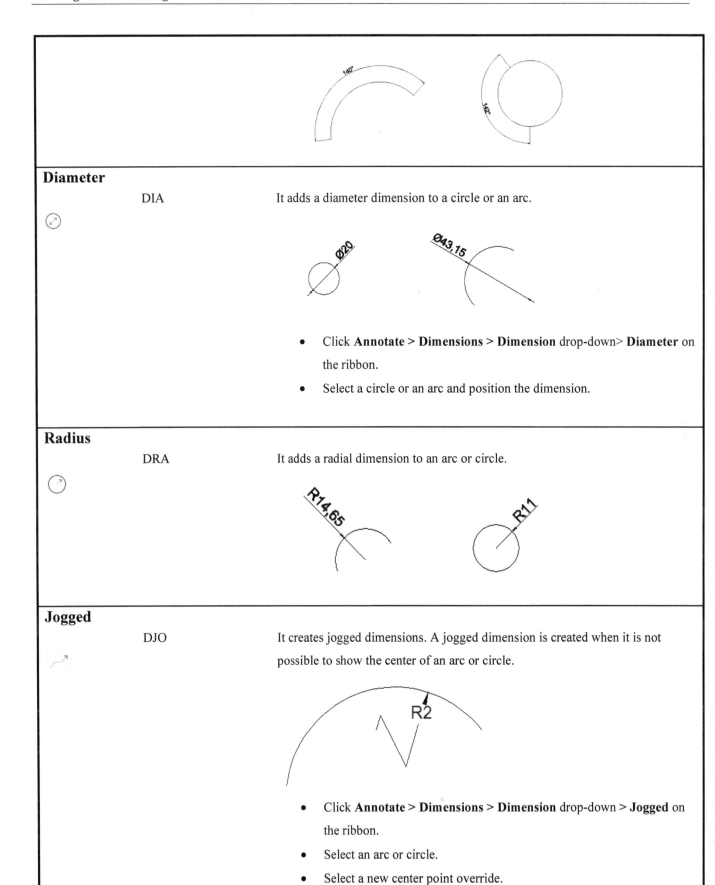

Diameter

DIA — It adds a diameter dimension to a circle or an arc.

- Click **Annotate > Dimensions > Dimension** drop-down> **Diameter** on the ribbon.
- Select a circle or an arc and position the dimension.

Radius

DRA — It adds a radial dimension to an arc or circle.

Jogged

DJO — It creates jogged dimensions. A jogged dimension is created when it is not possible to show the center of an arc or circle.

- Click **Annotate > Dimensions > Dimension** drop-down > **Jogged** on the ribbon.
- Select an arc or circle.
- Select a new center point override.

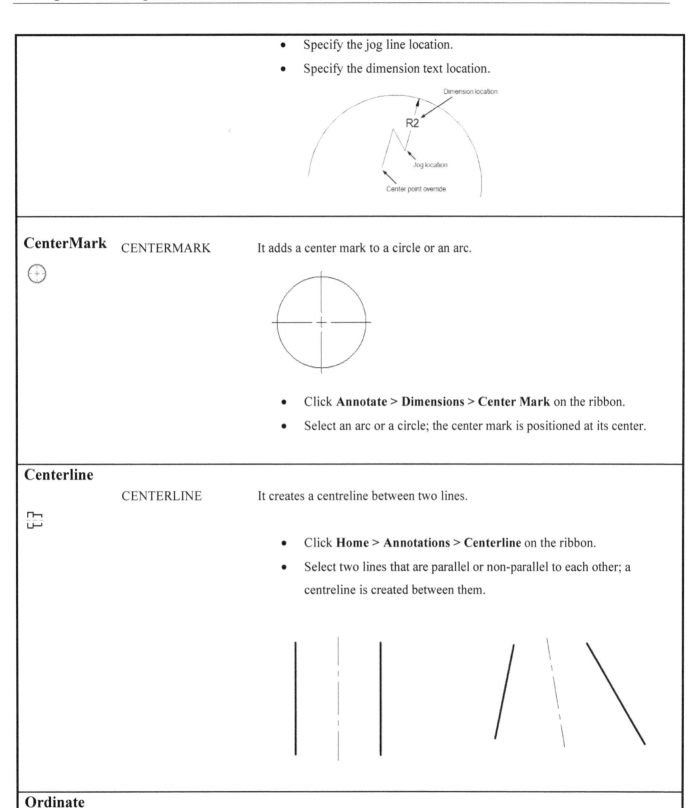

		• Specify the jog line location. • Specify the dimension text location.
CenterMark	CENTERMARK	It adds a center mark to a circle or an arc. • Click **Annotate > Dimensions > Center Mark** on the ribbon. • Select an arc or a circle; the center mark is positioned at its center.
Centerline	CENTERLINE	It creates a centreline between two lines. • Click **Home > Annotations > Centerline** on the ribbon. • Select two lines that are parallel or non-parallel to each other; a centreline is created between them.
Ordinate	DOR	It creates ordinate dimensions based on the current position of the User Coordinate System (UCS).

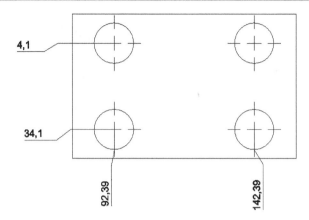

- Click **Annotate > Dimensions > Dimension** drop-down > **Ordinate** on the ribbon.
- Select the point of the object.
- Move the pointer in the vertical direction and click to position the X-Coordinate value.
- Select the point of the object.
- Move the pointer in the horizontal direction and click to position the Y-Coordinate value.

Split Dimension	DIMBREAK	It adds breaks to a dimension, extension, and leader lines.

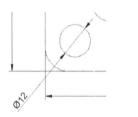 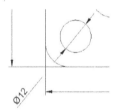

- Click **Annotate > Dimensions > Split Dimension** on the ribbon.
- Select the dimension to add a split.
- Select the dimension or object intersecting the dimension selected in the previous step. This splits the dimension by the intersecting object.
- Right-click and select ENTER.

Example:

In this example, you create the drawing, as shown in the figure, and add dimensions to it.

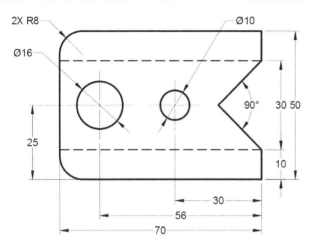

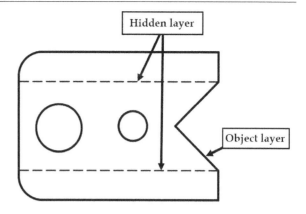

- Create four new layers with the following settings.

Layer	Lineweight	Linetype
Construction	0.00 mm	Continuous
Object	0.50 mm	Continuous
Hidden	0.30 mm	HIDDEN
Dimensions	Default	Continuous

- Type **LIMMAX**(GETXYURBNDS) and press ENTER.
- Type 100, 100, and press ENTER to set the maximum limit of the drawing.
- Click **View > Navigate > Zoom Window** drop-down >**Zoom Bounds** on the ribbon.
- Create the drawing on the **Object** and **Hidden** layers.

- Select the **Dimensions** layer from the **Layer Manager** drop-down in the **Dimensions** panel.

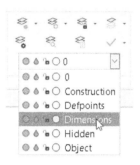

Creating a Dimension Style

The appearance of the dimensions depends on the dimension style that you use. You can create a new dimension style using the **Options** dialog. In this dialog, you can specify various settings related to the appearance and behavior of dimensions. The following example helps you to create a dimension style.

- Click the **Annotate > Dimensions > Dimension Style** icon on the ribbon.

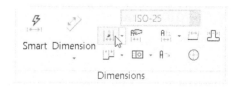

The **Options – Drafting Styles** dialog appears.

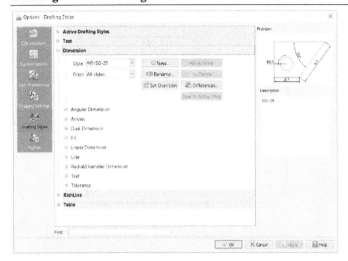

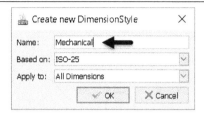

The basic nomenclature of dimensions is given below.

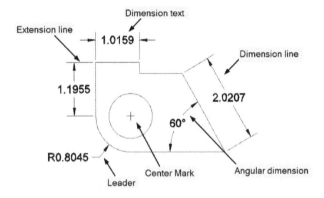

By default, the **ISO-25** or the **Standard** dimension style is active, depending upon the template that you have selected. If the default dimension style does not suit the dimensioning requirement, you can create a new dimension style and modify the nomenclature of the dimensions.

- To create a new dimension style, click the **New** ![New…] button on the **Options – Drafting Styles** dialog; the **Create new DimensionStyle** dialog appears.
- In the **Create new DimensionStyle** dialog, enter **Mechanical** in the **Name** box.
- Select **ISO-25** from the **Based on** drop-down and click **OK**.

- In the **Options- Drafting Styles** dialog, expand the **Linear Dimension** section.
- Ensure that the **Format** is set to **Decimal**.
- Set **Precision** to **0**.
- Select **Decimal separator** > '**.**'(Period).

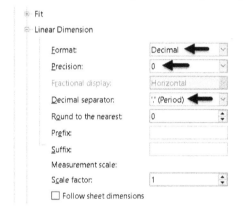

Study the other options in the **Linear Dimension** section. Most of them are self-explanatory.

- Expand the **Text > Text Settings** section.
- Ensure that the **Height** is set **2.5**.
- Expand the **Text Position** section under the **Text** section.
- Set the **Horizontal** and **Vertical** values to **Centered**.
- Expand the **Text alignment** section and select **Align Horizontally**.

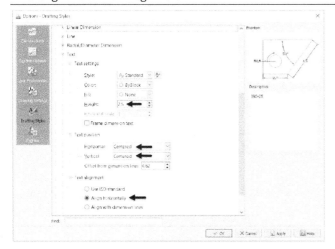

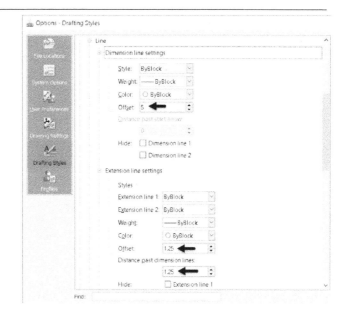

Study the other options under the **Text** section. These options let you change the appearance of the dimension text.

- Expand the **Line** section on the dialog.
- Under the **Line** section, expand the **Extension line settings** section.
- Set the **Offset** and **Distance past dimension lines** values to **1.25**.
- Expand the **Dimension line settings** section under the **Lines** section.
- Under the **Dimension line settings** section, set the **Offset** value to **5**.

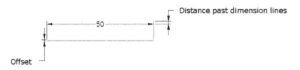

Study the different options in this section. The options in this tab are used to change the appearance and behavior of the dimension lines and extension lines.

- Expand the **Dimensions > Arrows** section and set **Size** to 3.
- Expand the **Radial/Diameter Dimension > Center mark display** section.
- Make sure that the **As Centerline** is selected and set **Size** to **3**.

Notice the different options in this section. The options in the **Dimension** section are used to change the appearance of the arrows and symbols. Also, you can set the appearance of the center marks and centrelines of circles and arcs.

- Click **OK** to accept the settings.

- On the **Annotate** tab, select **Mechanical** from the **Dimensions Style Control** drop-down available on the **Dimensions** group.

- On the **Annotate** tab of the ribbon, click **Dimensions > CenterMark** ⊕ .
- Select the circles from the drawing to apply the center mark to them.

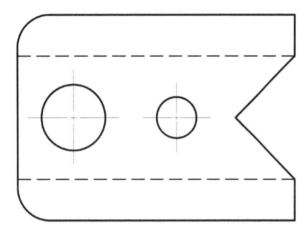

- On the ribbon, click **Annotate > Dimensions > Dimensions** drop-down **> Linear Dimension**.
- Make sure that the **ESnap** icon is turned on the status bar.
- Select the lower right corner of the drawing.
- Select the endpoint of the center mark of the small circle; the dimension is attached to the pointer.
- Move the pointer vertical downwards and position the dimension, as shown below.

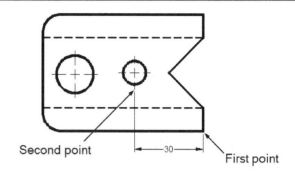

- Click the **Annotate > Dimensions > Continue** drop-down **> Baseline** on the ribbon; a dimension is attached to the pointer.
- Select the endpoint center mark of the large circle; another dimension is attached to the pointer.
- Select the lower-left corner of the drawing.
- Press ENTER twice.

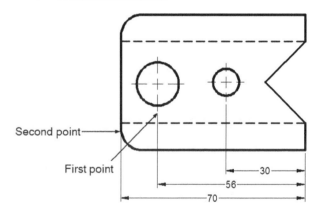

- Click the **Dimensions > Dimension** drop-down **> Angular** on the **Annotate** tab of the ribbon.
- Select the two angled lines of the drawing and position the angle dimension.

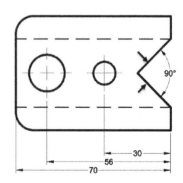

- Click the **Dimensions > Dimension** drop-down
  > **Diameter** ⊙ on the **Annotate** tab of the ribbon.
- Select the large circle and position the diameter dimension.
- Likewise, select the small circle and position the dimension.
- Click the **Dimensions > Dimension** drop-down
  > **Radius** ⊙ command on the **Annotate** tab of the ribbon.
- Select the fillet located at the top left corner; the radial dimension is attached to the pointer.
- Next, position the radial dimension approximately at 45 degrees.
- Select the radius dimension and click on the dimension palette icon; the **Dimension** palette appears.

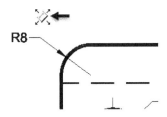

- In the Dimension palette, click in the **Prefix text** box and type 2X. Next, press the SPACEBAR.

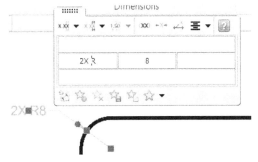

- Press Esc to deselect the radius dimension.
- Likewise, apply the other dimensions, as shown.
- Save and close the drawing.

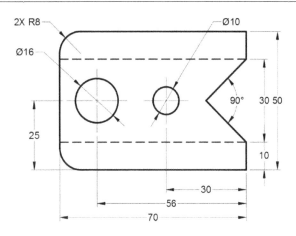

Adding Leaders

A leader is a thin solid line terminating with an arrowhead at one end and a dimension, note, or symbol at the other end. In the following example, you learn to create a leader style and then create a leader.

Example 1:

- Draw a square of 24 mm side length.
- Create a circle of 10.11 mm diameter at the center of the square.

- Click **Home > Draw > Arc > Center, Start, Angle** ⊙ on the ribbon.
- Select the center point of the circle.
- Move the pointer horizontally toward the right.
- Type 6 as the radius and press TAB.
- Type 0 and press ENTER to define the start angle of the arc.
- Type 270 as the included angle and press ENTER.

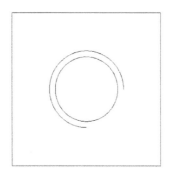

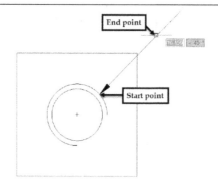

- Click the right mouse button on the **Polar (F10)** button on the status bar and select **Settings** from the menu.
- Set the **Incremental angle for Polar guide display** value to **45** degrees.
- Check the **Enable Polar guides** option and click **OK**.

- Click **Annotate > Dimensions > Smart Leader** on the ribbon.

- Select a point in the first quadrant of the arc.
- Move the pointer in the top-right direction and click to create the leader.

- Press ENTER twice.
- Type **M12x1.75 – 6H 16** in the text editor.

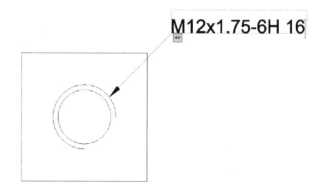

Next, you must insert the depth symbol before 16.

- Position the pointer before 16 and click the Insert **Symbol** drop-down on the **Note Formating** window.

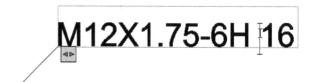

- Click **Other** on the menu; the **Character Map** dialog appears.

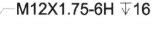

- In the **Character Map** dialog, select **GDT** from the **Font** drop-down.
- Select the Depth symbol from the fonts table.

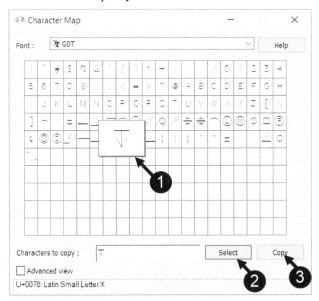

- Click **Select** and **Copy** buttons.
- Close the **Character Map** dialog.
- Right-click and select **Paste**; the depth symbol is pasted in the text editor.
- Adjust the spacing so that the complete text is in one line.
- Click in the graphics window.

Adding Dimensional Tolerances

During the manufacturing process, the accuracy of a part is an important factor. However, it is impossible to manufacture a part with the exact dimensions. Therefore, while applying dimensions to a drawing, we provide some dimensional tolerances, which lie within acceptable limits. The following example shows you to add dimensional tolerances in DraftSight.

Example:

- Create the drawing, as shown below. Do not add dimensions to it.

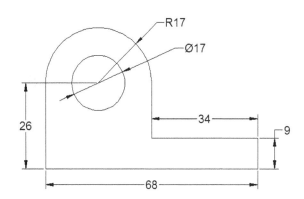

- Click the **Annotate** tab > **Dimensions** > **Dimension Style** on the ribbon.
- On the **Options-Drafting Styles** dialog, click the **New** button and type **Tolerances** in the **Name** box.
- Click **OK** on the **Create new DimensionStyle** dialog.
- Expand the **Tolerances** section under the **Dimension** section.

- Under the **Tolerances** section, select the **Calculation > Deviation** option.
- Set **Precision** as **0.00**.
- Set the **Maximum Value** and **Minimum Value** to **0.05**.
- Set the **Vertical text justification** as **Middle**.
- Specify the following settings accordingly:

Linear Dimension:

Format: Decimal

Precision: 0.00

Decimal Separator: '.'Period

Text:

Text Settings:

    Height: 2.5

Text position:

    Horizontal: Centered

    Vertical : Centered

Text alignment:

    Align horizontally

Arrows:

Size: 2.5

Radial/Diameter Dimension:

Center mark display:

    As Centerline

Size: 2.5

- Click **Activate** ⇨ Activate button on the **Options-Drafting Styles** dialog.
- Click **OK** to close the dialog
- Apply dimensions to the drawing.

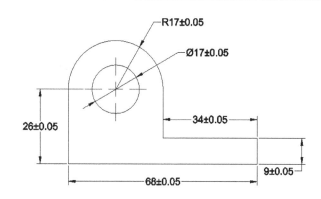

Geometric Dimensioning and Tolerancing

Earlier, you have learned how to apply tolerance to the size (dimensions) of a component. However, the dimensional tolerances are not sufficient for manufacturing a component. You must give tolerance values to its shape, orientation, and position, as well. The following figure shows a note which is used to explain the tolerance value given to the shape of the object.

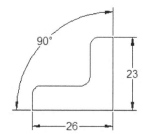

Providing a note in a drawing may be confusing. To avoid this, we use Geometric Dimensioning and Tolerancing (GD&T) symbols to specify the tolerance values to shape, orientation, and position of a component. The following figure shows the same example represented by using the GD&T symbols. In this figure, the vertical face to which the tolerance frame is connected must be within two parallel planes 0.08 apart and perpendicular to the datum reference (horizontal plane).

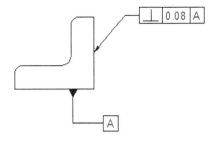

The Geometric Tolerancing symbols that can be used to interpret the geometric conditions are given in the table below.

Purpose		Symbol
To represent the shape of a single feature.	Straightness	
	Flatness	
	Cylindricity	
	Circularity	
	Profile of a surface	
	Profile of a line	
To represent the orientation of a feature with respect to another feature.	Parallelism	
	Perpendicularity	
	Angularity	

To represent the position of a feature with respect to another feature.	Position	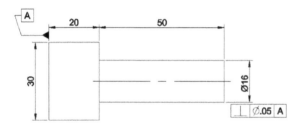
	Concentricity and coaxiality	
	Run-out	
	Total Run-out	
	Symmetry	

Note: table symbols shown in images

Example 1:

In this example, you apply geometric tolerances to the drawing shown below.

- Create the drawing, as shown below.

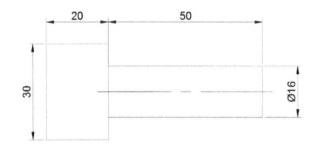

- Click **Annotate > Dimensions > Tolerance** on the ribbon; the **Geometric Tolerance** dialog appears.

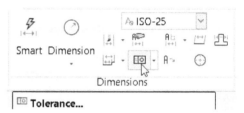

- In the **Geometric Tolerance** dialog, click the **Symbol** drop-down > **Perpendicular** symbol.

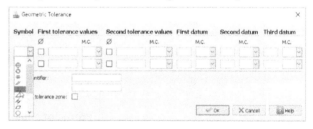

- Select the checkbox under the diameter symbol of the **First tolerance values** section.

- Enter **.05** in the box next to the checkbox under the **First tolerance values** section.

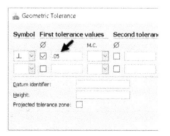

- Enter **A** in the upper box of the **First datum** section.

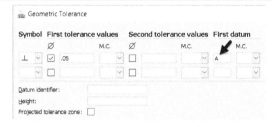

- Click **OK** and position the **Feature Control frame**, as shown below.

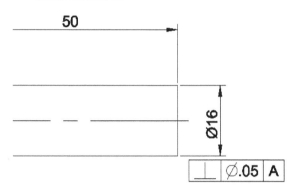

Next, you must add the datum reference.

- On the **Annotate** ribbon tab, click **Dimensions** panel > **Smart Leader** .
- Type S and press ENTER.
- On the **Format Leaders** dialog, click the **Annotations** tab.
- Select **Type > Tolerance**.
- Click the **Arrows/Lines** tab.
- Select **Arrow Style > Datum triangle filled**.
- Click **OK**.
- Specify the first, second, and third points of the datum reference, as shown.

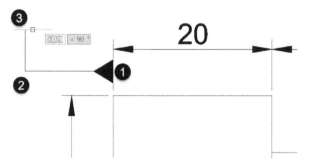

- Press ENTER.

- On the **Geometric Tolerance** dialog, type **A** in the **Datum Identifier** box.

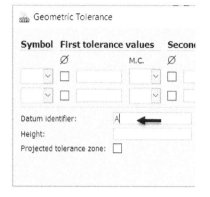

- Click **OK**.

Editing Dimensions by Stretching

In DraftSight, the dimensions are associative to the drawing. If you modify a drawing, the dimensions are modified automatically. In the following example, you stretch the drawing to modify the dimensions.

Example:

- Create the drawing, as shown below, and apply dimensions to it.

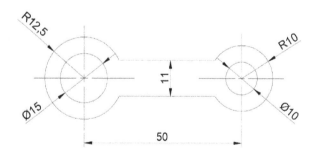

- Click **Home > Modify > Stretch** on the ribbon.
- Drag a window and select the right-side circle, arc, and horizontal lines.

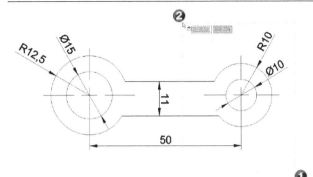

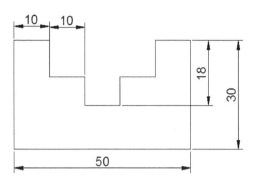

dimensions to it.

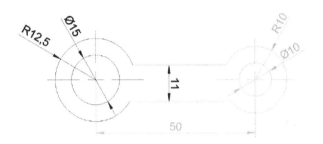

- Right-click and select the center point of the right-side circle.

- Move the pointer to stretch the drawing; notice that the horizontal dimension also changes.

- Type **30** and press ENTER; the horizontal dimension is updated to 80.

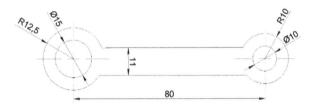

Modifying Dimensions by Trimming and Extending

In earlier chapters, you have learned to modify drawings by trimming and extending objects. In the same way, you can modify dimensions by trimming and extending. The following example shows you to modify dimensions by this method.

Example:

- Create a drawing, as shown below, and add

- Click **Home** > **Modify** > **Trim** drop-down > **Trim** ✂ on the ribbon.

- Select the horizontal line to define the cutting edge, as shown.

- Right-click and press Enter to accept.

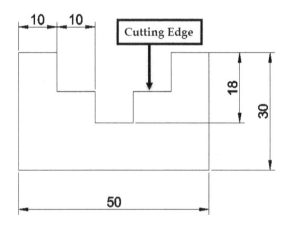

- Select the vertical dimension with the value 18 and press **Delete** on the keyboard.

- Add the dimension, as shown in the figure.

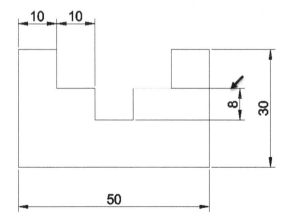

- Press ESC.

- Click **Home > Modify > Trim > Extend** ⊤ on the ribbon.

- Select the vertical edge as the boundary, as shown below. Next, right-click to accept.

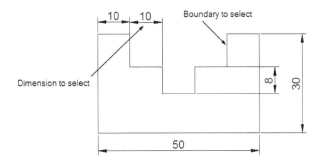

- Select the horizontal dimension with the value 10. The dimension is extended up to the selected boundary.

- Press Esc.

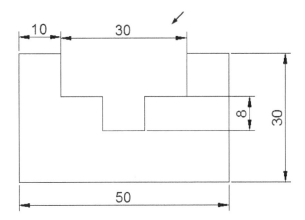

Using the EDITDIMENSION command

The EDITDIMENSION command can be used to modify dimensions. Using this command, you can add text to a dimension, rotate the dimension text and extension lines, or reset the position of the dimension text.

Example 1: (Adding Text to the dimension)

- Type EDITDIMENSION in the command line and press ENTER.

- Type **N** or **New** in the command line. Next, press ENTER; a text box appears.

- Enter **TYP** in the command line and press the SPACEBAR and type 10.

- Press Enter and select the dimension with value 10.

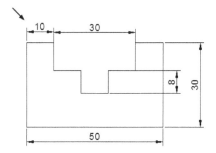

- Press ENTER; the dimension text is changed.

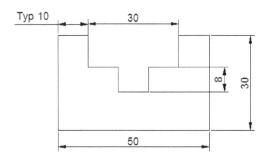

Example 2: (Rotating the dimension text)

- Enter **DED** in the command line.

- Type A or Angle in the command line and then press ENTER; the message, "Specify new text angle," appears in the command line.

- Type **30** and press ENTER.

- Select the dimension with the value 50 and right-click. The angle of the dimension text is changed to 30 degrees. Note that the angle is measured from the horizontal axis (X-axis).

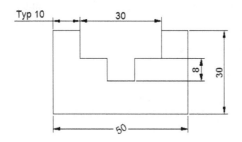

Using the Oblique tool

The **Oblique** tool is used to incline the extension lines of a dimension. This tool is very useful while dimensioning the isometric drawings. It can also be used in 2D drawings when the dimensions overlap with each other.

Example:

In this example, you create an isometric drawing and add dimensions to it. Next, you use the **Oblique** tool to change the angle of the dimension lines.

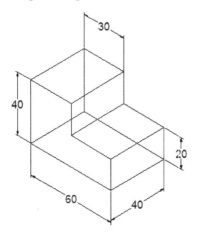

- Click the right mouse button on the **Grid** icon, and then select **Settings**.
- Select the **Orientation > Isometric** option under the **Grid Settings** section.

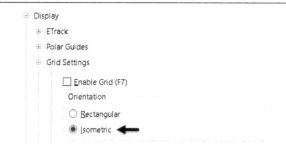

- Click **OK**.
- Turn on the **Snap** and the **Ortho** . Also, turn on the **QInput** .
- Click **Navigate > Zoom Window** drop-down > **Zoom Fit** on the **View** tab of the ribbon.
- Type **L** in the command line and press ENTER.
- Click at a random point and move the pointer vertically.
- Type 40 in the command line and press ENTER; a vertical line is created.
- Move the pointer toward the right; notice that an inclined line is attached to the pointer.

- Type 30 and press ENTER; an inclined line is drawn.
- Move the pointer downward.
- Type 20 and press ENTER.

- Move the pointer toward the right.

- Type 30 and press ENTER.

- Move downward, type 20, and then press ENTER.

- Move the pointer toward the left and click on the start point of the sketch.

- Right-click select **Enter**.

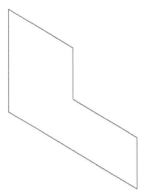

- Turn off the **Ortho**.

- Create a selection window and select all the objects of the sketch.

- Right-click and select **Copy** from the shortcut menu.

- Select the lower-left corner point as the base point.

- Move the pointer toward the right.

- Type 40 and press TAB.

- Type 45 and press ENTER.

- Right-click and select **Enter**.

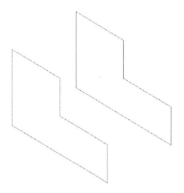

- Use the **Line** tool and connect the endpoints of the two sketches.

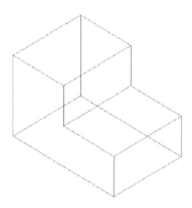

- Click the right mouse button on the **Grid** icon available at the status bar.

- Select **Settings** from the menu.

- Expand the **Display > Grid Settings** section.

- Select the **Orientation > Rectangular** option and click **OK**.

- Deactivate the **Snap** icon on the Status bar.

- Use the dimensioning tools and apply dimensions to the sketch.

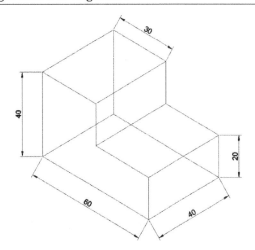

- Click the **Annotate** tab > **Dimensions** panel > **Oblique** button on the ribbon.

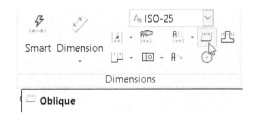

- Select the vertical dimensions and right-click to accept; the message, "**Specify oblique angle**," appears in the command line.

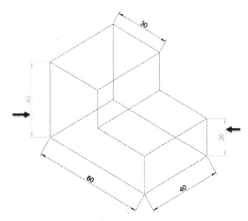

- Type 150 as the oblique angle and press ENTER; the dimensions are oblique, as shown below.

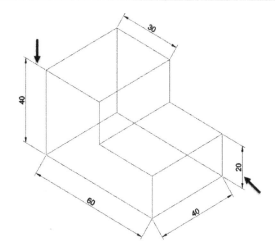

- Again, click the **Oblique** tool on the **Dimensions** panel and select the aligned dimensions. Next, right-click to accept.

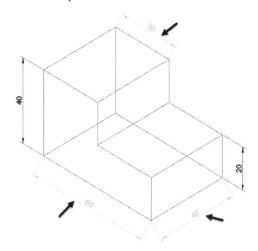

- Type 90 as the oblique angle and press ENTER; the dimensions are oblique, as shown below.

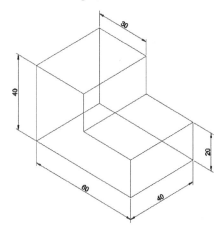

Editing Dimensions using Grips

In Chapter 4, you have learned to edit objects using grips. In the same way, you can edit dimensions using grips. The editing operations using grips are discussed next.

Example 1: (Stretching the Dimension)

- Select the dimension to display grips on it.
- Select the endpoint grip of the dimension.
- Next, move the pointer and select a new point; the dimension value is updated automatically.

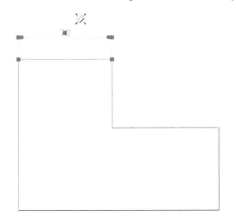

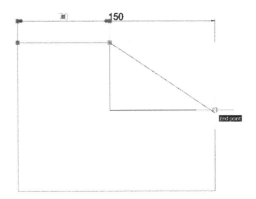

- You can also stretch angular or radial dimensions.

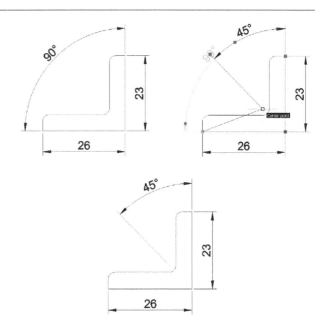

Example 2: (Moving the Dimension)

- To move a linear dimension, select the middle grip and move the pointer.

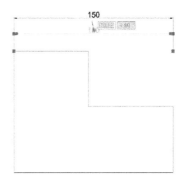

- Likewise, you can move the angular and radial dimensions.

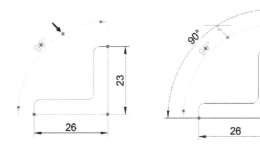

Modifying Dimensions using the Properties palette

Using the **Properties** palette, you can modify the dimensional properties such as text, arrow size, precision, linetype, lineweight, and so on. The **Properties** palette comes in handy when you want to modify the properties of a particular dimension only.

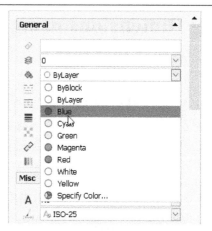

Example:

- Create the drawing shown in the figure and apply dimensions to it.

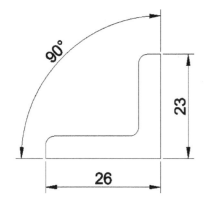

- Select the vertical dimension and right-click.
- In the **Properties** palette, under the **Lines & Arrows** section, set the **Arrow size** to **2**.

- Under the **General** section, set **LineColor** to **Blue**.

- In the **Properties** palette, under the **Lines & Arrows** section, set the **Ext line offset** value to **1.25**.
- Scroll down to the **Text** section and set **Text height** to **2**.

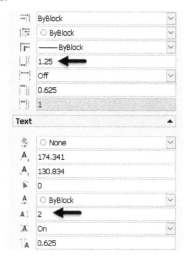

- Press Esc to deselect the dimension; the properties of the dimension are updated as per the changes made.

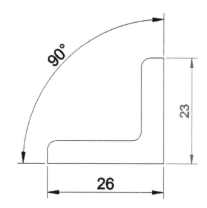

Matching Properties of Dimensions or Objects

In the previous section, you have learned to change the properties of a dimension. Now, you can apply these properties to other dimensions by using the **Properties Painter** tool.

- Click **Home > Properties > Properties Painter** on the ribbon or type **MA** and press ENTER; the message, "Specify source entity>>" appears in the command line.

- Select the vertical dimension from the drawing; the message, "Options: Settings or Specify destination entities>>," appears in the command line.

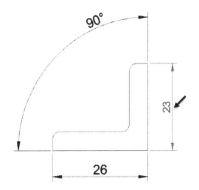

- Type **S** or **Settings** in the command line and press ENTER; the **Property Painter** dialog appears.

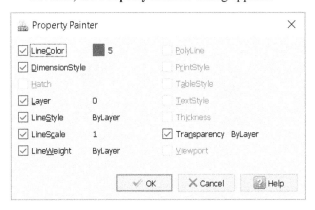

In this dialog, you can select the settings that can be applied to the destination dimensions or objects. By default, all the options are selected in this dialog.

- Click **OK** on the **Property Painter** dialog. Next, you must select the destination objects.

- Select the other dimensions from the drawing; the properties of the source dimension are applied to other dimensions.

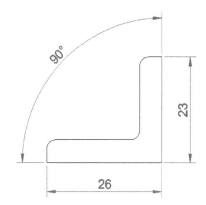

- Right-click and select **Enter**.

Exercises

Exercise 1

Create the drawing shown below and create hole callouts for different types of holes. Assume missing dimensions.

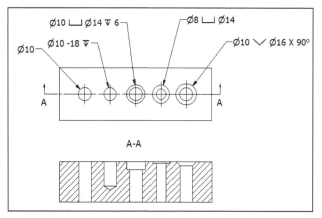

Exercise 2

Create the following drawings and apply dimensions and annotations. The Grid Spacing X= 10 and Grid Spacing Y=10.

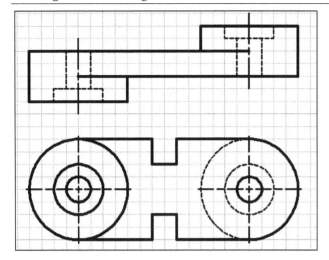

Precision: 0.00

Upper Value: 0.05

Lower Value: 0.05

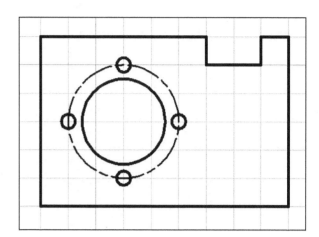

Exercise 3

Create the drawing shown below. The Grid spacing is 10 mm. After creating the drawing, apply dimensional tolerances to it. The tolerance specifications are given below.

Method: Limits

Chapter 7: Section Views

In this chapter, you learn to:

- **Create Section Views**
- **Set Hatch Properties**
- **Use Island Detection tools**
- **Create text in Hatching**
- **Edit Hatching**

Section Views

In this chapter, you learn to create section views. You can create section views to display the interior portion of a component that cannot be shown clearly by means of hidden lines. This can be done by cutting the component using an imaginary plane. In a section view, section lines, or cross-hatch lines are added to indicate the surfaces that are cut by the imaginary cutting plane. In DraftSight, you can add these section lines or cross-hatch lines using the **Hatch** tool.

The Hatch tool

The **Hatch** tool is used to generate hatch lines by clicking inside a closed area. When you click inside a closed area, a temporarily closed boundary is created using the **Polyline** command. The closed boundary is filled with hatch lines, and then it is deleted.

Example 1:

In this example, you apply hatch lines to the drawing, as shown in the figure below.

- Open a new DraftSight file.

- Create four layers with the following properties.

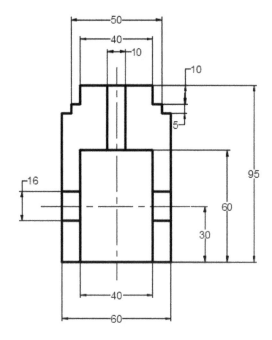

- Create the drawing, as shown below. Do not apply dimensions.

- Select the **Hatch lines** layer from the **Layer Manager** drop-down of the **Layers** group.
- Click **Home > Draw > Hatch** on the ribbon, or enter **H** in the command line; the **Hatch Fill** dialog appears on the screen.

- On the **Hatch/Fill** dialog, click the **Preview Patterns** button next to **Pattern** drop-down under the **Pattern** section, as shown.

- Select **ANSI31** on the **Select Pattern Style** dialog.

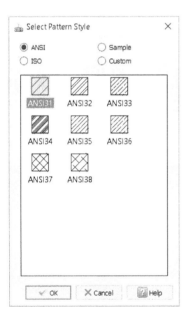

- Click **OK**.
- Click the **Specify points** button under the **Boundary settings** section.

- Select the boundaries of the four regions, as shown.
- Right click and select ENTER.

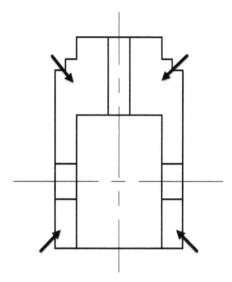

- Click the **OK** button on the **Hatch/Fill** dialog.

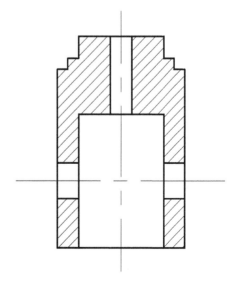

Example 2:

In this example, you create the front and section views of a crank.

- Create five layers with the following settings:

Layer	Lineweight	Linetype
Construction	0.00 mm	Continuous
Object	0.30 mm	Continuous
Centerline	0.00 mm	CENTER
Hatch lines	0.00 mm	Continuous
Cutting Plane	0.30 mm	PHANTOM

- Activate the **Construction** layer and create construction lines, as shown.

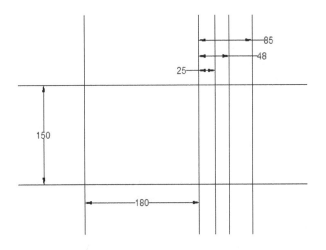

- Set the **Object** layer as current and create draw circles, as shown below.

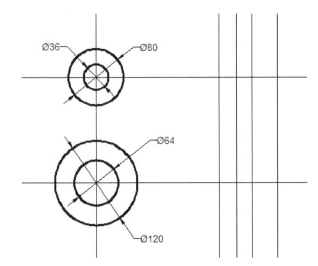

- Switch to the **Construction** layer and create construction lines, as shown.

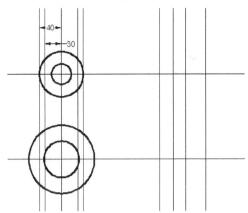

- Switch to **Object** layer and create two lines, as shown.

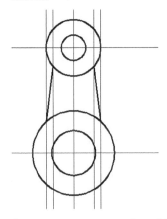

- On your own, create other objects on the front view, as shown. The construction lines are hidden in the image.

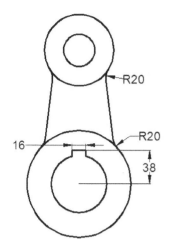

- On your own, create the objects of the section view, as shown below. (For any help, refer to the **Multi view Drawings** section of Chapter 5)

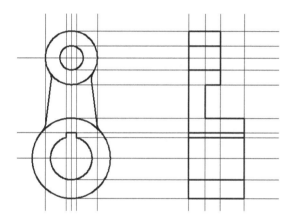

- Set the **Centerlines** layer as current and create center marks and centrelines.

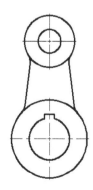 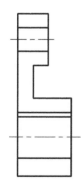

- Set the **Cutting Plane** layer as current.
- Click the **Polyline** button on the **Draw** panel and pick a point below the front view, as shown.

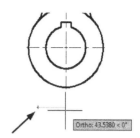

- Type **W** or **Width** in the command line.
- Press ENTER.
- Type 0 as the starting width and press ENTER.
- Type 10 as the ending width and press ENTER.
- Move the pointer horizontally toward the right and enter 20.
- Again, type **W** or **Width** in the command line. Next, press ENTER.
- Type 0 as the starting width and press ENTER.
- Type 10 as the ending width and press ENTER.
- Move the pointer and place it on the endpoint of the center mark, as shown.

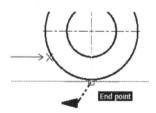

- Move the pointer downward and click when trace lines are displayed, as shown below.

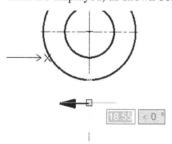

- Move the pointer vertically up and click.
- Move the pointer to the endpoint arrow.

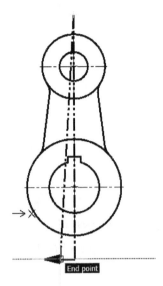

- Move the pointer upward and click when trace lines are displayed from the endpoint of the lower horizontal line.

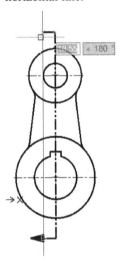

- Type **W** or **Width** in the command line.
- Press ENTER.
- Type 10 as the starting width and press ENTER.
- Type 0 as the ending width and press ENTER.
- Move the pointer horizontally toward left and enter 20.

- Activate the **Hatch lines** layer.
- Type **H** in the command line and press ENTER; the **Hatch/Fill** dialog appears.
- Click the **Preview patterns** button next to the **Pattern** drop-down; the **Select Pattern Style** dialog appears.

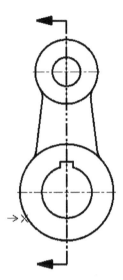

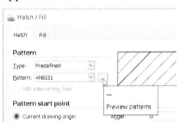

- Select the **ANSI** option, and then select **ANSI31** from the dialog.
- Click **OK**.
- Set the **Scale** value to **2**.

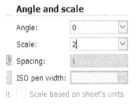

- Click the **Specify Points** button in the **Boundary Settings** section and click in Region 1, Region 2 and Region 3.

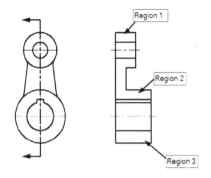

- Press ENTER to create hatch lines.

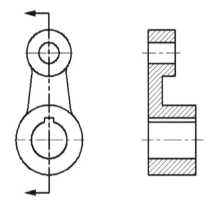

- Click **OK** on the **Hatch/Fill** dialog.
- Save the drawing as **Crank.dwg** and close.

Setting the Properties of Hatch lines

You can set the properties of the hatch lines such as angle and scale in the **Angle and Scale** section of the **Hatch/Fill** dialog. You can also set some additional properties such as transparency, origin, and internal regions by clicking the **Additional Options** button on the **Hatch/Fill** dialog. These properties are available on the **Additional Options-Area Hatch/Fill** dialog.

Example:

- Create four layers with the following settings.

Layer	Lineweight	Linetype	LineColor
Construction	0.00 mm	Continuous	White
Object	0.30 mm	Continuous	White
Centerline	0.00 mm	CENTER	White
Hatch lines	0.00 mm	Continuous	Blue

- Create the following drawing in different layers. Do not apply dimensions.

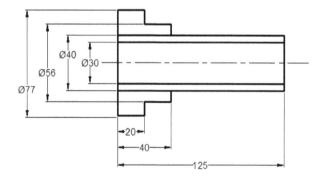

- Type **H** and press ENTER; the **Hatch / Fill** dialog appears on the screen.
- Click the **Type > Predefined** option under the **Pattern** section on the dialog.

You can also select a different hatch type, such as Custom and User-defined.

- Select **ANSI31** from the **Pattern** drop-down.
- On the **Additional Options – Area Hatch/Fill** dialog, click the **Layer** drop-down and select **Hatch lines**.
- Make sure that the transparency is **0** under the **Transparency** section.
- Click **OK** on the **Additional Options – Area Hatch/Fill** dialog.
- Click the **Specify points** button on the **Hatch/Fill** dialog under the **Boundary settings** section.

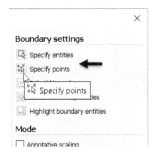

- Pick points in the outer areas of the drawing, as shown below.

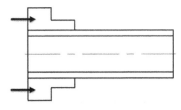

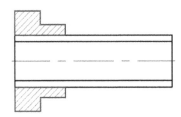

- Press Enter to display the **Hatch / Fill** dialog.
- Adjust the **Scale** to **1.5** under the **Angle and Scale** section.
- Click the **Preview** button; the distance between the hatch lines changes.

- Press ESC.
- Click the **OK** button on the **Hatch/Fill** dialog to close.
- Press the SPACEBAR to activate the **HATCH** command again.
- Change the **Angle** value to **90** under the **Angle and Scale** section.

- Click the **Specify points** button on the **Hatch/Fill** dialog under the **Boundary settings** section.
- Pick points in the area, as shown below.

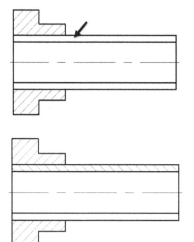

On zooming into the hatch lines, you may notice that they are not aligned properly. This is because the **Current drawing origin** is selected under the **Pattern start point** section on the **Hatch/Fill** dialog. As a result, the origin of the drawing acts as the origin of the hatch pattern.

However, you can change the origin of the hatch pattern.

- On the **Hatch/Fill** dialog, select the **User-defined location** option under the **Pattern start point** section.

- Click the **Select in graphics area** button, as shown.

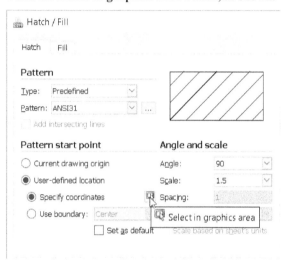

- Next, select the origin point, as shown below. Then, press Enter.

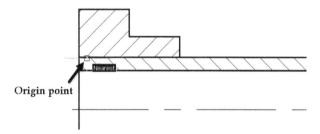

Origin point

- Click **OK** on the **Hatch/Fill** dialog.
- Likewise, create another hatch, as shown.

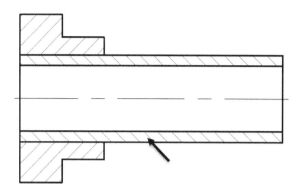

- Save and close the file.

Island Detection tools

While creating hatch lines, the island detection tools help you to detect the internal areas of a drawing.

Example:

- Create the drawing, as shown below. Do not apply dimensions.

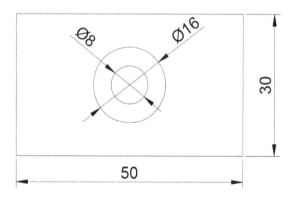

- Click **Home > Draw > Hatch** on the ribbon.
- Select **ANSI31** from the **Pattern** drop-down on the **Hatch/Fill** dialog.
- Click the **Additional Options** button on the dialog.
- Select the **In/Out** option under the **Internal regions** section on the **Additional Options-Area Hatch/Fill** dialog.

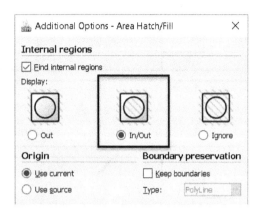

- Click **OK** on the **Additional Options – Area Hatch/Fill** dialog.

- On the **Hatch/Fill** dialog, click the **Specify points** icon under the **Boundary settings** section.

- Pick a point in the area outside the large circle; the area inside the small circle is detected automatically. Also, hatch lines are created inside the small circle.

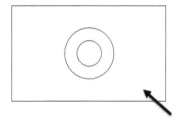

- Press **Enter** and click **OK** on the **Hatch/Fill** dialog.

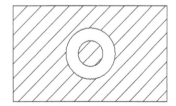

- Click **Undo** on the **Quick Access Toolbar**.

- Activate the **Hatch** tool and select **ANSI31** from the **Pattern** drop-down on the **Hatch/Fill** dialog.

- Click the **Additional Options** button on the dialog.

- Select the **Out** option under the **Internal regions** section on the **Additional Options-Area Hatch/Fill** dialog.

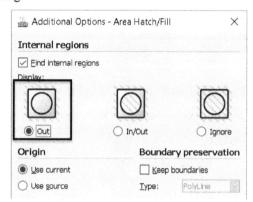

- Click **OK** on this dialog and click the **Specify points** icon under the **Boundary settings** section on the **Hatch/Fill** dialog.

- Pick a point in the area outside the large circle and press ENTER; hatch lines are created only outside the large circle. The **Out** tool enables you to create hatch lines only in the outermost level of the drawing.

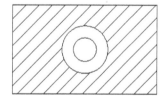

- Repeat the process using the **Ignore** option. The internal loops are ignored while creating the hatch lines.

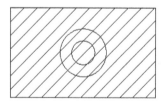

Text in Hatching

You can create hatching without passing through the text and dimensions.

- Create a drawing, as shown in the figure.

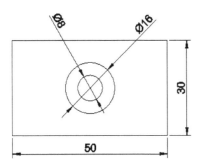

- Apply a dimension to the large circle.
- Click **Home** > **Annotations** > **Text** drop-down > **Note** on the ribbon.

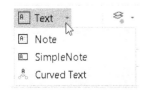

- Specify the first and second corners of the text editor, as shown below.

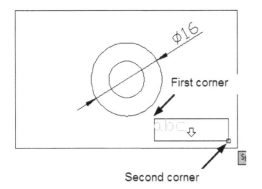

- Select **Arial** from the **Font** drop-down on the **Note Formatting** dialog.

- Ensure that **Text Height** is set to **2.5**.

- Type **DraftSight** in the text editor. Left-click in the empty space of the graphics window.

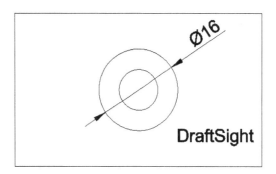

- Activate the **Hatch** tool and click the **Additional Options** button on the **Hatch/Fill** dialog.
- Select the **In/Out** option from the **Internal regions** section on the **Additional Options** dialog.
- Click **OK** on this dialog and click the **Specify points** icon under the **Boundary settings** section on the **Hatch/Fill** dialog.
- Pick a point in the area covered by the outside boundary and press ENTER. Click **OK** on the **Hatch/ Fill** dialog; hatch lines are created. The hatch lines do not pass through the text and dimension.

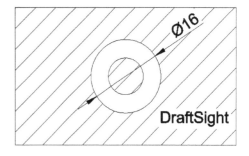

Editing Hatch lines

You can edit a hatch by using the **Edit Hatch** tool or simply selecting the hatch.

- To edit a hatch, select the hatch lines from the drawing and right-click.
- Select the **Hatch Edit** option from the shortcut menu. You can also type **EDITHATCH** in the command line and select the hatch from the drawing.

- The **Hatch/Fill** dialog appears.
- Edit the options in the **Hatch/Fill** dialog and click the **OK** button; the hatch pattern is modified.

Exercises

Exercise 1

Create the half-section view of the object shown below.

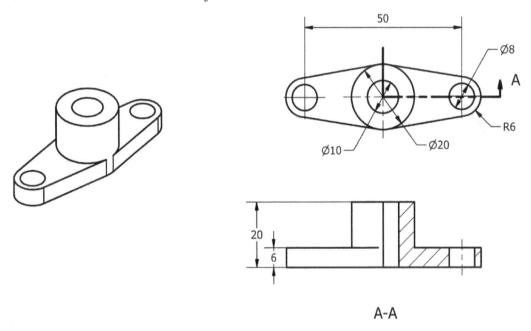

A-A

Exercise 2

In this exercise, the top, front, and right side views of an object are given. Replace the front view with a section view. The section plane is given in the top view.

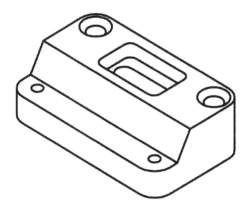

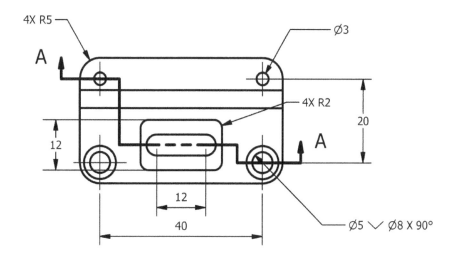

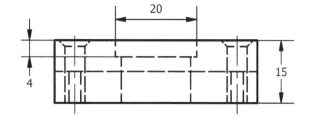

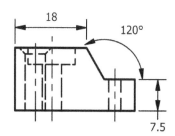

Chapter 8: Blocks, Attributes, and References

In this chapter, you learn to do the following:

- **Create and insert Blocks**
- **Create Annotative Blocks**
- **Explode and purge Blocks**
- **Use the Divide tool**
- **Use the DesignCenter and Tool Palettes to insert Blocks**
- **Insert Multiple Blocks**
- **Edit Blocks**
- **Create Blocks using the Write Block tool**
- **Define and insert Attributes**
- **Work with References**

Introduction

In this chapter, you learn to create and insert Blocks and Attributes in a drawing. You will also learn to attach external references to a drawing. The first part of this chapter deals with Blocks. A Block is a group of objects combined and saved together. You can later insert it in drawings. The second part of this chapter deals with Attributes. An Attribute is an intelligent text attached to a block. It can be any information related to the block, such as description, part name, and value. The third part of the chapter deals with the References (external references). External references are drawing files, images, PDF files attached to a drawing.

Creating Blocks

To create a block, first, you need to create shapes using the drawing tools and use the BLOCK command to convert all the objects into a single object. The following example shows the procedure to create a block.

Example 1

- Create the drawing, as shown below. Do not apply dimensions. Assume the missing dimensions.

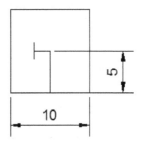

- Click **Insert > Block Definition > Define Block** on the ribbon; the **Block Definition** dialog appears.

- Enter **Target** in the **Name** field.
- Click the **Select in graphics area** button under the **Entities** section on the dialog. Drag a window and select all the objects of the drawing.

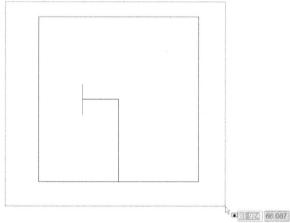

- Right-click to accept; the dialog appears again. You can choose to retain or delete the objects after defining the block. The **Preserve as separate entities** option under the **Entities** section retains the objects in the graphics window after defining the block. The **Convert to block** option deletes the objects and displays the block in place of them. The **Remove from drawing** option completely deletes the objects from the graphics window.

- Select the **Remove from drawing** option under the **Entities** section.

- Click the **Select in graphics area** button under the **Base point** on the dialog.

- Select the midpoint of the left vertical line. The selected point is the insertion point when you insert this block into a drawing.

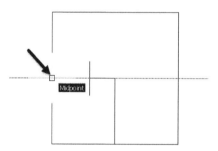

You can also add a description to the block in the **Description** box. In addition to that, you can set the behavior of the block such as scalability, annotative, and explode ability using the options in the **Settings** section. The options in the **Settings** area can be used to set the units of the block and link a website or other files with the block.

- Uncheck the **Apply uniform scale** option (for this example).

- Click **OK** on the dialog; the block is created and saved in the database.

Inserting Blocks

After creating a block, you can insert it at the desired location inside the drawing using the INSERT command.

The procedures to insert blocks are explained in the following examples.

Example 1

- Click **Insert > Block > Insert Block** on the ribbon; the **Insert Block** dialog appears on the screen.

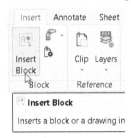

- On the dialog, select the **Target** block from the **Name** drop-down.

- Check the **Specify later** option under the **Position** section.

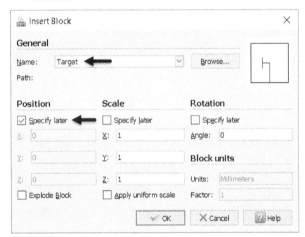

- Click **OK**.
- Pick a point in the graphics window to place the block.

Example 2 (Scaling the block)

- On the ribbon, click **Insert > Block > Insert Block**; the **Insert Block** dialog appears.

- Select **Target** from the **Name** drop-down.
  You can use the options in the **Scale** section to scale the block. The **Apply uniform scale** option can be used to scale the block uniformly. You can uncheck this option to specify the scale factor separately in the

X, Y, and Z boxes. If you check the **Specify later** option, the block can be scaled dynamically in the graphics window.

- Check the **Specify later** option under the **Position** section.

- Check the **Specify later** option in the **Scale** section.

- Uncheck the **Apply uniform scale** option in the **Scale** section.

- Click **OK**; the block is attached to the pointer.

- Pick a point in the graphics window; the message, "Options: Corner, uniform scale or Specify X scale factor or specify opposite corner >>" appears in the command line. In addition, as you move the pointer, the block automatically scales.

- Type 3 and press ENTER; the message, "Specify Y-scale >>" appears.

  - Type 2 as the Y scale factor and press ENTER; the block is scaled, as shown below.

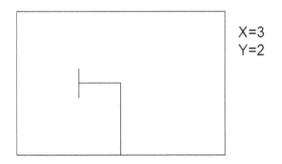

Example 3 (Rotating the block)

- Click **Home > Block > Insert Block >** on the ribbon; the **Insert Block** dialog appears.

- Select **Target** from the **Name** drop-down.

- Check the **Specify later** option under the **Position** section.

- Uncheck the **Specify later** option in the **Scale** section and check the **Apply uniform scale** option.
  The options in the **Rotation** section are used to rotate the block. You can enter the rotation angle in the

Angle box. You can dynamically rotate the block by selecting the **Specify later** option.

- Check the **Specify later** option in the **Rotation** section.

- Click **OK** and pick a point in the graphics window; the message, "Specify angle >>:" appears in the command line. As you rotate the pointer, the block also rotates. You can dynamically rotate the block and pick a point to orient the block at an angle or type a value and press ENTER to specify the angle.

- Type **45** and press **ENTER**; the block is rotated by **45** degrees.

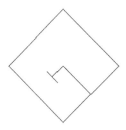

- Save and close the drawing file.

Creating Annotative Blocks

Annotative blocks possess annotative properties. They are scaled automatically depending upon the scale of the drawing sheet. The procedure to create and insert annotative blocks is explained in the following example.

Example:

- Create the drawing, as shown in the figure. Assume the missing dimensions.

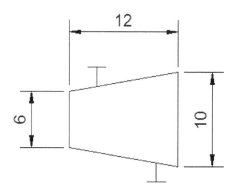

- Click **Insert > Block Definition > Define Block** on the ribbon; the **Block Definition** dialog appears.

- Enter **Turbine Driver** in the **Name** field.

- Click the **Select in graphics area** button under the **Entities** section on the dialog. Create a window and select all the objects of the drawing. Right-click to accept the selection.

- Select the **Remove from drawing** option under the **Entities** section.

- Click the **Select in graphics area** button under the **Base point** section and select the midpoint of the left vertical line.

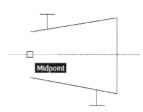

- Check the **Annotative scaling** option under the **Settings** section. Click the **OK** button on the dialog.

- On the Status bar, click the **Annotation** drop-down and select 1:10.

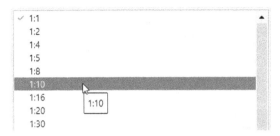

- Click **Insert > Block > Insert Block**.

- Uncheck the **Specify Later** option under the **Rotation** section.

- Click **Name > Turbine Driver** on the **Insert** dialog.

- Check the **Specify Later** option under the **Position** section.

- Click **OK**.

- Pick a point in the graphics window; the block is inserted with the scale factor 1:10.

- On the ribbon, click **View** tab > **Navigate** group > **Zoom** drop-down > **Zoom Bounds** to view the block.

- Change the **Annotation Scale** to **1:2**.

- Select the block and then right click.

- Select **Annotative Entity Scale > Add Current Scale**; the block is automatically scaled to **1:2**.

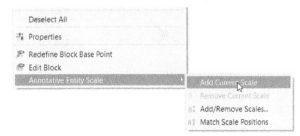

Exploding Blocks

When you insert a block into a drawing, it is considered as a single object, even though it consists of numerous individual objects. At many times, you may require to break a block into its individual parts. Use the **Explode** tool to break a block into its individual objects.

- To explode a block, click **Home > Modify > Explode** on the ribbon or type EXPLODE in the command line and press ENTER.

- Select the block and press ENTER; the block is broken into individual objects. You can select the individual objects by clicking on them. (Refer to the Explode Tool section of Chapter 4).

Using the Clean command

You can remove the unused blocks and other unwanted drawing data from the database using the **Clean** command.

- To delete the unused data, type the **Clean** command in the command line; the **Clean** dialog appears.

- To remove unwanted blocks from the database, expand the **Blocks** tree, and select the blocks.
- Click the **Delete** button on the dialog.
- Click **Close** on the **Clean** dialog.

Using the Divide tool

The **Divide** tool is used to place a number of instances of an object equally spaced on a line segment. You can also place blocks on a line segment. The following example shows you to divide a line using the **Divide** tool.

Example:

- Create the object, as shown in the figure.

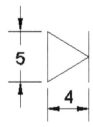

- Create a block with the name **Diode**. Specify the midpoint of the left vertical line as the base point.
- Create a line of 50 mm length and 45 degrees inclination.
- Expand the **Draw** panel in the **Home** tab and click the

Multiple Points drop-down > **Make Division**.

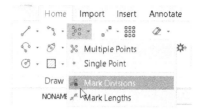

- Select the line segment; the message, "Options: Block or Specify number of segments >>" appears.
- Type **B** or **Block** in the command line; the message, "Specify block name>>" appears.
- Type **Diode** and press ENTER; the message, "Align block with entity? Specify Yes or No>>" appears.
- Type Y or Yes in the command line; the message, "Specify number of segments>>" appears.
- Type **5** and press ENTER; the line segment is divided into five segments, and four instances of blocks are placed.

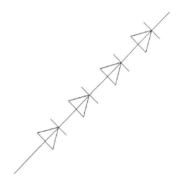

- Trim the unwanted portions, as shown below.

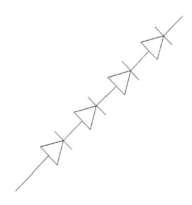

Renaming Blocks

You can rename blocks at any time. The procedure to rename blocks is discussed next.

- Type **RENAME** in the command line and press ENTER; the **Rename** dialog appears.
- In the **Rename** dialog, expand **Blocks** from the **All Items** list.
- Select the block to be named
- Enter a new name in the **Name** box.
- Click the **Rename** button; the block is renamed.

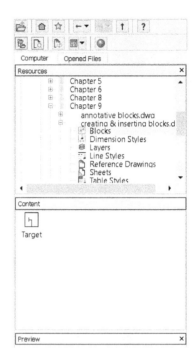

- Click **Close**.

Using the Design Resources

Design Resources is one of the additional means by which you can insert blocks and drawings in an effective way. Using the Design Resources, you can insert blocks created in one drawing into another drawing. You can display the Design Resources by clicking **Insert > Resources > Design Resources** on the ribbon (or) entering **DC** in the command line. The following example shows you to insert blocks using the Design Resources.

Example:

- Open a new drawing file.
- Create the following symbols and convert them into blocks. You can also download them from the companion website.

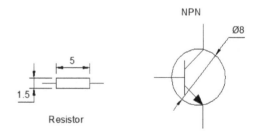

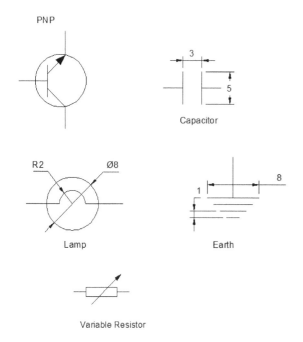

PNP

Capacitor

R2   Ø8

Lamp

Earth

Variable Resistor

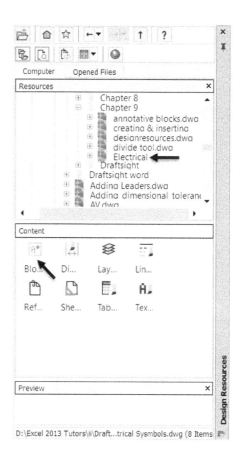

- Save the file as **Electronic Symbols.dwg**. Close the file.

- Open a new drawing file.

- Set the maximum limit of the drawing to 100,100.

- Click **View** tab > **Navigate** > **Zoom drop-down** > **Zoom Bound** on the ribbon.

- Click **Insert** > **Resources** > **Design Resources** on the ribbon; the **Design Resources** palette appears.

- In the **Design Resources** palette, browse to the location of the **Electronic Symbols.dwg** file using the Folder List. Select the file and double-click on the **Blocks** icon; all the blocks present in the file are displayed.

- Drag and place the blocks in the graphics window.

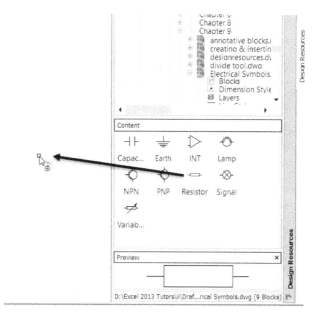

You can also insert blocks by activating the **Insert Block** dialog.

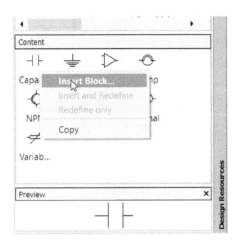

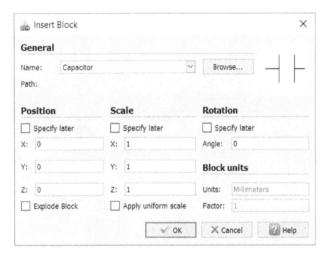

- Use the **Move** and **Rotate** tools and arrange the blocks, as shown below.

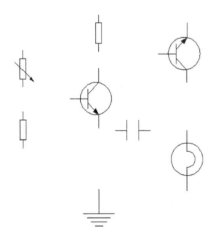

- Use the **Line** tool and complete the drawing, as shown below.

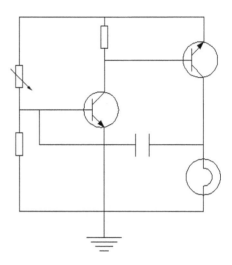

Inserting Multiple Blocks

You can insert multiple instances of a block at a time by using the **INSERTBLOCKN** command. This command is similar to the **PATTERN** command. The following example explains the procedure to insert multiple blocks at a time.

Example:

- Create two blocks, as shown below.

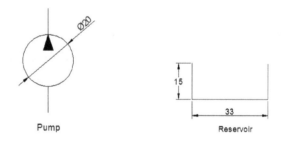

- Type **INSERTBLOCKN** in the command line and press ENTER; the message, "Specify block name>>" appears.
- Type **Pump** and press ENTER; the Pump is attached to the pointer.
- Pick a point in the graphics window.
- Type 1 as the scale factor.
- Press ENTER twice.

　　Blocks, Attributes, and References

- Enter 0 as the rotation angle; the message, "Specify number of horizontal rows," appears.

- Enter 1 as the row value; the message, "Specify number of vertical columns," appears.

- Enter 4 as the column value; the message, "Specify distance between columns (|||):" appears.

- Type 60 and press ENTER; the pumps are inserted, as shown below.

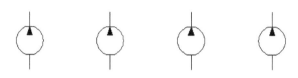

- Likewise, insert the reservoirs and create lines, as shown below.

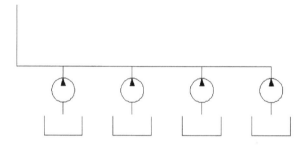

Editing Blocks

During the design process, you may need to edit blocks. You can easily edit a block using the **Edit Component** command. As you edit a block, all the instances of it are automatically updated. The procedure to edit a block is discussed next.

- Click **Insert > Component > Edit Component** on the ribbon.

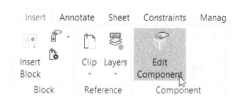

- Select **Pump** from the graphics window and click **OK**; the **EditComponent** window appears.

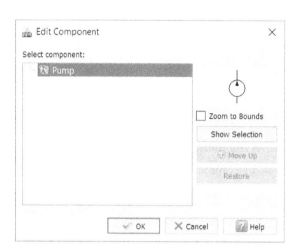

- Click **Home > Draw > Polyline** on the ribbon and draw a polyline, as shown below.

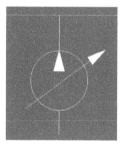

- Click **Insert > Component > Close Component** on the ribbon.

- In the **Component** dialog, click **Save**.

All the instances of the block are updated automatically.

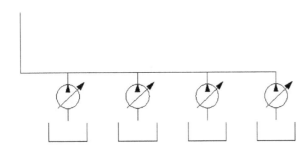

Using the Write Block tool

Using the **Write Block** tool, you can create a drawing file from a block or objects. You can later insert this drawing file as a block into another drawing. The procedure to create a drawing file using blocks is discussed in the following example.

Example:

- Start a new drawing file and create two blocks, as shown below. You can also download them from the companion website.

- Insert the blocks and create the drawing, as shown below.

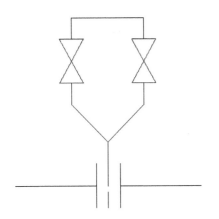

- Type **Setbase** in the command line, and then press ENTER.

- Select the endpoint of the lower horizontal line, as shown.

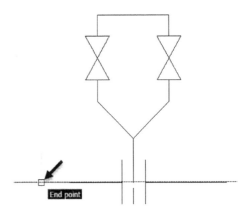

- Click **File > Export > Export Drawing**; the **Save File** dialog appears.

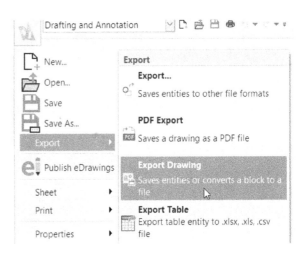

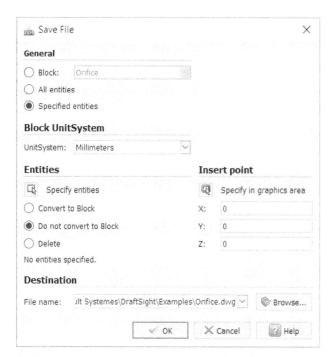

In the **Save File** dialog, you can select three different types of sources (Block, All entities, or Specified entities) to create a block. If you select the **Block** option, you can select blocks present in the drawing from the drop-down.

- Select the **All entities** option.
- Specify the location of the file and name it as **Tap-in line**.

- Click the **OK** button.
- Close the drawing file.
- Open a new drawing file, and then type **I** in the command line and press ENTER; the **Insert Block** dialog appears.
- Click the **Browse** button and go to the location of the **Tap-in Line** file.
- Select **Tap-in line** from the **Name** drop-down and click **OK**.
- Pick a point in the graphics window to insert the block.

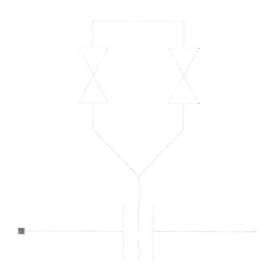

Defining Attributes

An attribute is a line of text attached to a block. It may contain any type of information related to a block. For example, the following image shows a Compressor symbol with an equipment tag. The procedure to create an attribute is discussed in the following example.

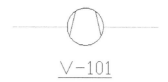

Example 1:

- Open a new drawing file.
- Create the symbols, as shown below.

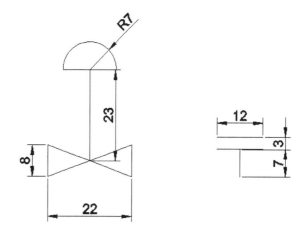

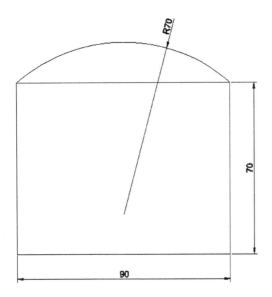

- Click **Insert > Block Definition > Define Block Attribute** on the ribbon; the **Block Attribute Definition** dialog appears.

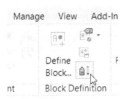

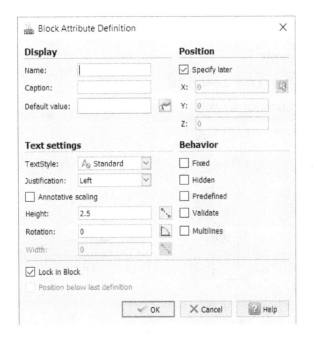

The options in the **Behavior** group of the **Block Attribute Definition** dialog define the display mode of the attribute. If you check the **Hidden** option, the attribute is invisible. The **Fixed** option makes the value of the attribute constant. You cannot change the value. The **Validate** option prompts you to verify after you enter a value. The **Predefined** option can be used to set a predefined value for the attribute. The **Lock in Block** option fixes the position of the attribute to a selected point. The **Multilines** option allows typing the attribute value in single or multiple lines.

- Ensure that the **Lock in Block** option is selected.

The options in the **Display** group define the values of the attribute. The **Name** box is used to enter the label of the attribute. For example, if you want to create an attribute called RESISTANCE, you must type **Resistance** in the **Tag** box. The **Caption** box defines the prompt message that appears after placing the block. The **Default value** box defines the default value of the attribute.

- In the Block **Attribute Definition** dialog, enter **Valvetag** in the **Name** box.

The **Text Settings** options define the display properties of the text, such as style, height, and so on. Observe the other options in this dialog. Most of them are self-explanatory.

- Enter **5** in the **Height** box.
- Set the **Justification** to **Middle** and click **OK.**
- Specify the location of the attribute, as shown below.

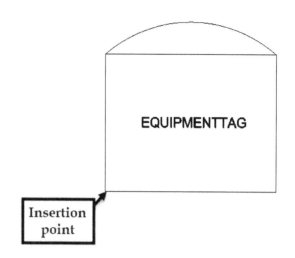

- Click the **Define Block** button on the **Block Definition** panel; the **Block Definition** dialog appears.
- On the dialog, click the **Select in graphics area** button in the **Entities** group.
- Drag a window to select the control valve symbol and attribute.

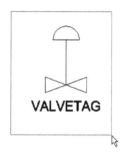

- Also, create a block of the nozzle symbol and name it **Nozzle**.

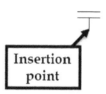

- Press ENTER.
- Select the **Remove from drawing** option from the **Entities** group.
- Click the **Select in graphics area** button under the **Base point** group and select the point, as shown below.

- Enter **Control Valve** in the **Name** box and click **OK**.
- Likewise, create the **Equipmenttag** attribute and place it inside the tank symbol.
- Create a block and name it as **Tank**.

Inserting Attributed Blocks

You can use the Insert Block command to insert the attributed blocks into a drawing. The procedure to insert attributed blocks is discussed next.

- On the ribbon, click **Insert > Block > Insert Block**.
- Select **Tank** from the **Name** drop-down.
- Check the **Specify later** option in the **Position** section.
- Click **OK**.
- Click in the graphics window to define the insertion point.
- Enter **TK-001** and press ENTER; the block is placed along with the attribute.

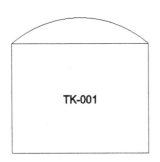

- Likewise, place control valves, as shown below.

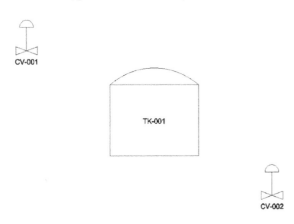

- Place the nozzles on the tank, as shown below.

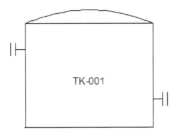

- Use the **Polyline** tool and connect the control valves and tank.

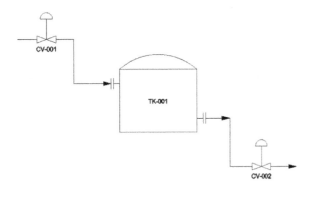

Working with External references

Attach

In DraftSight, you can attach a drawing file, image, or pdf file to another drawing. These attachments are called External References (References). They are dynamic in nature and update when changes are made to them. In the following example, you learn to attach drawing files to a drawing.

Example 1:

- Create the drawing shown below.

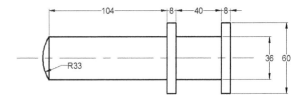

- Type **Setbase** in the command line and press ENTER.
- Select the midpoint of the vertical line as the base point.

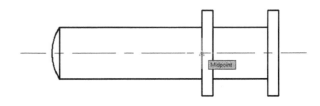

- Save the drawing as **Crank pin.dwg**
- Create another drawing as shown below (For help, refer to the **Multi View Drawings** section in Chapter 5).

Blocks, Attributes, and References

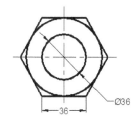

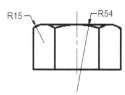

- Type **Setbase** in the command line and press ENTER.
- Specify the base point, as shown below.

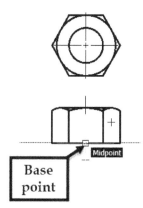

- Save the drawing as **Nut.dwg** and close it.
- Open the **Crank.dwg** file created in Chapter 8.
- On the ribbon, click **Insert** tab > **Blocks** panel > **References Manager**.

- In the **References** palette, click the **Attach Drawing** drop-down; the **Select file** dialog appears.

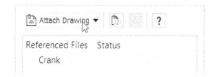

- Browse to the location of the **Crankpin.dwg** and double-click on it; the **Attach Reference:Drawing** dialog appears.

Some of the options available in this dialog are similar to that in the **Insert Block** dialog, such as the Position, Scale, and Rotation of the external reference.

- Check the **Specify later** option in the **Positon** group of this dialog and click **OK**; the crank pin is attached to the pointer.
- Select the point on the hatched view, as shown below.

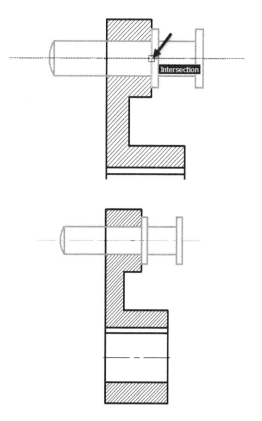

- In the **References** palette, select the Crank file.

- Click the **Attach Drawing** drop-down; the **Select file** dialog appears.

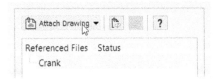

- Browse to the location of the **Nut.dwg** and double-click on it; the **Attach Reference: Drawing** dialog appears.

- In the **Attach Reference: Drawing** dialog, enter **90** in the **Angle** box under the **Rotation** group.

- Check the **Specify later** option in the **Positon** group of this dialog and click **OK**.

- Select the insertion point on the section view, as shown below.

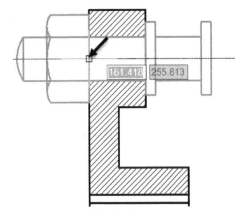

Clipping External References

You can hide the unwanted portion of an external reference by using the **Clip Reference** tool.

- Click **Insert > Reference > Clip Reference** on the ribbon; the message, "Specify references," appears in the command line.

- Select the **Nut.dwg** from the graphics window.

- Press ENTER

- Type **B** in the command line.

- Press ENTER.

- Type **R** in the command line.

- Press ENTER.

- Draw a rectangle as shown below; only the front view of the nut is visible, and the top view is hidden. Also, the clipping frame is visible.

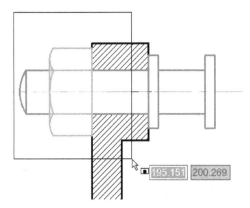

- Attach another instance of the **Nut.dwg** file.

- Use the **Rotate** and **Move** tools to position the top view, as shown below.

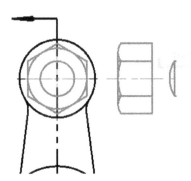

- Use the **Clip Reference** tool and clip the Xref.

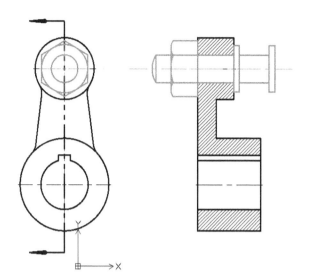

Editing the External References

DraftSight allows you to edit the external references in the file to which they are attached. You can also edit them by opening their drawing files. The procedure to edit an external reference is discussed next.

- Select the reference, and then click the right mouse button.
- Select the **Edit Reference** option from the shortcut menu.

- Select **Nut** from the **Edit Component** dialog.
- Click **OK** to get into the reference editing mode.
- In the drawing, the centerlines of the nut are overlapping on the centerlines of the crank. Delete the centerlines and center marks of the nut.
- Right click and select **Close Component**; the **Component** message box appears.
- Click **Save**.

Adding Balloons

- Create a new layer and name it as Annotation.
- Set the **Linestyle** to **Continuous Solid line**.
- Set the **Lineweight** to 0.
- Make the layer as current.
- Create a circle of 24 mm in diameter.
- On the ribbon, click **Insert > Block Definition > Define Block Definition**.
- Type **Balloon** in the **Name** box.
- Select **Justification > Middle**.
- Type **10** in the **Height** box.
- Click **OK**.
- Select the center point of the circle.

- Click **Insert > Block Definition > Define Block** on the ribbon.
- On the **Block Definition** dialog, type **Balloon** in the **Name** box.
- Click the **Select in graphics area** icon in the **Entities** group.
- Create a selection window across the circle and attribute.
- Click the **Select in graphics area** icon in the **Position** group.
- Select the left quadrant point of the circle.

- Select the **Remove from drawing** option in the **Entities** group.

- Click **OK**.
- Click **Insert > Block > Insert Block** on the ribbon.
- Select the **Balloon** block from the **Name** drop-down.
- Check the **Specify Later** option in the **Position** group.
- Click **OK**.
- Click in the graphics area to place the block.
- Type 1 and press ENTER.

- Click **Insert > Block > Insert Block** on the ribbon.
- Select the **Balloon** block from the **Name** drop-down.
- Check the **Specify Later** option in the **Position** group.
- Click **OK**.
- Click in the graphics area to place the block.
- Type 2 and press ENTER.

- Click **Insert > Block > Insert Block** on the ribbon.
- Select the **Balloon** block from the **Name** drop-down.
- Check the **Specify Later** option in the **Position** group.
- Click **OK**.
- Click in the graphics area to place the block.
- Type 3 and press ENTER.

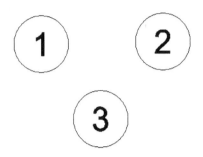

- Click **Annotate > Dimensions > Dimension style** on the ribbon.
- On the **Drafting Styles** dialog, expand the **Arrows** node and type **8** in the **Size** box.

- Click **Apply**.
- Click the User Preferences option on the left side of the dialog.
- On the dialog, expand **Drafting Options > Display > Polar Guides**.
- Type **45** in the **Incremental angles for Polar guide display** box.
- Check the **Enable Polar guides (Polar)** option.
- Click **OK**.
- Click **Annotate > Dimensions > Smart Leader** on the ribbon.
- Type S in the command line and press ENTER; the **Format Leaders** dialog appears.
- Click the **Annotation** tab on the dialog.
- Select **Type > Copy entity**.
- Click **OK**.
- Select a point on the section view of the crank.
- Move the pointer along the polar trace lines and click.

- Move the pointer horizontally toward the right and click.

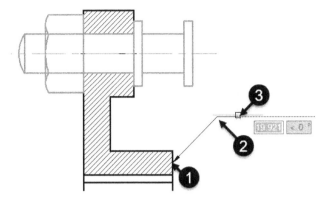

- Select the first balloon from the graphics area.

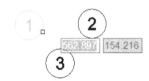

The balloon is created.

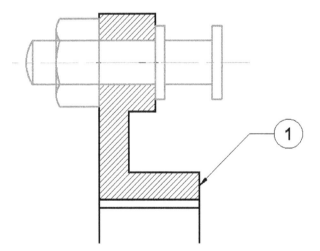

- Likewise, create other balloons.
- Delete the blocks from the graphics area.

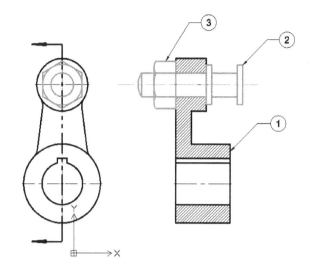

Creating Part List

- Click **Annotate > Table > Table Style** on the ribbon; the **Drafting Styles** dialog appears.

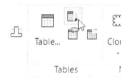

- In the **Drafting Styles** dialog, click the **New** button; the **Create New TableStyle** dialog appears,
- In the **Create New Table Style** dialog, enter **Part List** in the **Name** box.
- Click **OK**.
- In the **Text** section, set the **Height** to **10**.

- Select the **Head** option from the **Contents** drop-down in the **Cell Styles settings** section.
- In the **Text** section, set the **Height** to **10**.
- Click **OK**.
- Click the **Table** button on the **Tables** panel; the **Insert table** dialog appears.

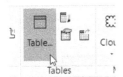

- Ensure that the **Table style** is set to **Part List**.

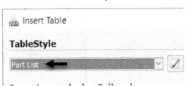

- Under the **Cell Styles** group, set the **First row** to **Header**.

- Set the **Second row** and **All other row** to **Data**.

- In the **Columns** group, type **4** and **65** in the **Number** and **Width** boxes, respectively.

- In the **Rows** group, type **2** and **1** in the **Number** and **Height** boxes, respectively.

- Click **OK** and place the table at the lower right corner of the graphics window.

- Enter **PART No.**, **NAME**, **MATERIAL**, **QTY** in the first row of the **Part list** table. Use the **TAB** key to navigate between the cells.

- Click **Create Note** button on the **Note Formatting** panel.

- Click on any one of the edges of the table; the grips are displayed on it. You can edit the table using these grips.

- Click and drag the square grip below the MATERIAL cell; the width of the cell is changed.

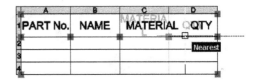

- Likewise, adjust the widths of the remaining cells, as shown.

PART No.	NAME	MATERIAL	QTY

- Click and drag the grip located at the bottom left corner of the table; the height of the rows is increased uniformly.

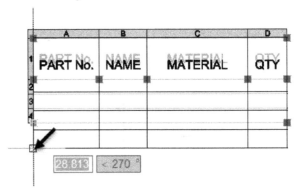

- Click in the second cell of the first column; the **Table** ribbon appears. You can use this ribbon to modify the properties of the table cell.

- Click the **Row Below** button on the **Insert** panel; a new row is added to the cell.

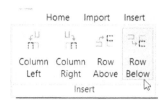

PART No.	NAME	MATERIAL	QTY
1	Crank	Forged Steel	1
2	Crank pin	45C	1
3	Nut	MS	1

- Double-click in the cell below the **PART No.**; the text editor is activated.

- Enter the following data in the cells. Use the TAB to navigate between the cells.

Exercise

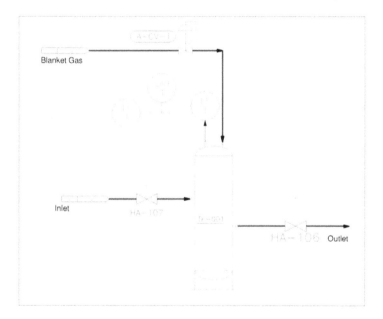

Chapter 9: Sheets & Annotative Objects

In this chapter, you learn to do the following:

- **Create Sheets**
- **Specify the Paper space settings**
- **Create Viewports in Paper space**
- **Change Layer properties in Viewports**
- **Create a Title Block on the Sheet**
- **Use Annotative objects in Viewports**

Drawing Sheets

There are two workspaces in DraftSight: The Model space and the Paper space. In the Model space, you create drawings. You can even plot drawings from the model space. However, it is difficult to plot drawings at a scale or if a drawing consists of multiple views arranged at different scales. For this purpose, we use Sheets or paper space. In Sheets or paper space, you can work on notes and annotations and perform the plotting or publishing operations. In Sheets, you can arrange a single view or multiple views of a drawing or multiple drawings by using Viewports. These viewports display drawings at specific scales on Sheets. They are mainly rectangular in shape, but you can also create circular and polygonal viewports. In this chapter, you learn about viewports and various annotative objects.

Working with Sheets

Sheets represent the conventional drawing sheet. They are created to plot a drawing on a paper or in electronic form. A drawing can have multiple Sheets to print in different

sheet formats. By default, there are two Sheets available: Sheet1 and Sheet2. You can also create new Sheets by clicking the plus (+) symbol next to the sheet. Next, select **New Sheet** from the shortcut menu. In the following example, you create two Sheets, one representing the ISO A1 (841 X 594) sheet and another representing the ISO A4 (210 X 297) sheet.

Example:

- Open a new drawing file.
- Create layers with the following settings:

Layer	Linetype	Lineweight
Construction	Continuous	Default
Object Lines	Continuous	0.6mm
Hidden Lines	Hidden	0.3 mm
Center Lines	CENTER	Default
Dimensions	Continuous	Default
Title Block	Continuous	1.2mm
Viewport	Continuous	Default

- Create the drawing, as shown next. Do not add dimensions.

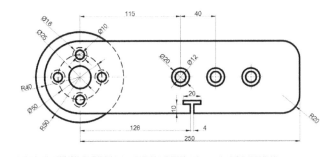

- Click the **Sheet 1** tab at the bottom of the graphics window.

You notice that a white paper is displayed with the viewport created automatically. The components of a Sheet are shown in the figure below.

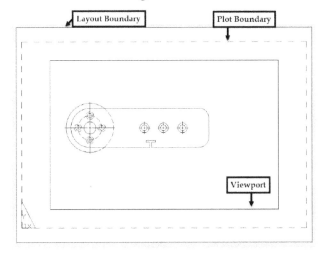

- Click the right mouse button on the **Sheet 1** tab and select **Print Configuration Manager**; the **Print Configuration Manager** dialog appears.

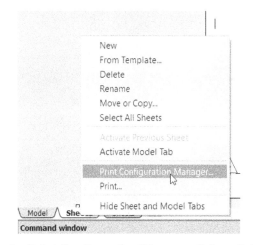

- In the **Print Configuration Manager** dialog, click the **New** button; the **New Print Configuration** dialog appears.

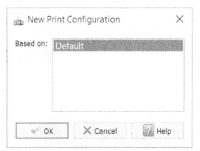

- Select the **Default** option and click **OK**.
- Type **ISO A1** in the **File name** box
- Click **Save**.
- On the **Print Configuration** dialog, select **PDF** from the **Name** drop-down under the **Printer/Plotter** group. 9
- Set the **Paper size** to **ISO A1 (841.00 x 594.00 MM)**.
- Select **1:1** from the drop-down available in the **Scale** section.
- Set the **Orientation** to **Landscape**.

- Click the **Additional Options** button.
- Set the **PrintStyle table** to **default.stb**.

- Click **OK**, and then click **Save** on the **Print Configuration** dialog.
- Select the **ISO A1** option.
- Click **Activate**.
- Click **Close** on the **Print Configuration Manager** dialog.
- Double-click on the **Sheet1** tab and enter **ISO A1**; the **Sheet1** is renamed.

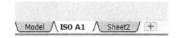

- Similarly, rename the **Sheet2** to **ISO A4**.

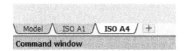

- Click the right mouse button on the **ISO A4** tab and select **Print Configuration Manager**; the **Print Configuration Manager** dialog appears.

- Select **ISO A1** from the list.
- Click the **New** button.
- Select the **ISO A1** option and click **OK**.
- Type **ISO A4** in the **File name** box
- In the **Print Configuration** dialog, select the **PDF** plotter.
- Set the **Paper size** to **ISO A4 (210 x 297 MM)** and **Scale** to **1:1.**
- Set **Drawing Orientation** as **Portrait** and click **Save**; the size of the Sheet is changed to A4 size.

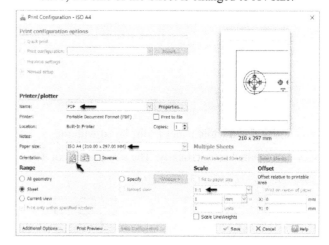

- Select the **ISO A4** option.
- Click **Activate**.
- Close the **Print Configuration Manager** dialog.

Creating Viewports in the Paper space

The viewports that exist in the paper space are called floating viewports. This is because you can position them anywhere in the Sheet and modify their shape size with respect to the Sheet.

Creating a Viewport in the ISO A4 Sheet

- Open the **ISO A4** Sheet, if not already open.
- Select the default viewport that exists in the **ISO A4** Sheet.
- Press the DELETE key; the viewport is deleted.
- Click **Sheet > Viewports > Single Viewport** on the

ribbon.

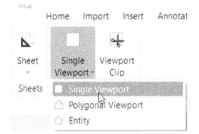

- Create the rectangular viewport by picking the first and second corner points, as shown in the figure.

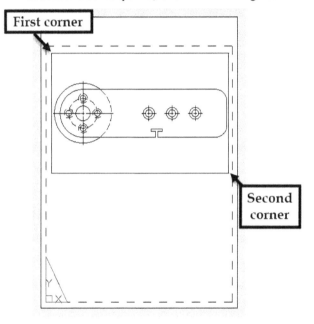

First corner

Second corner

- Double-click inside the viewport; the model space inside the viewport is activated. In addition, the viewport frame becomes thicker when you are in model space.

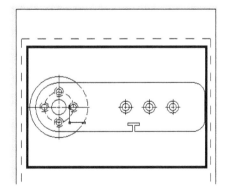

- Click the **Annotation** drop-down on the status bar, and then select **1:2** from the menu; the drawing is zoomed out.

- After fitting the drawing inside the viewport, you can lock the position by double-clicking outside the Viewport.

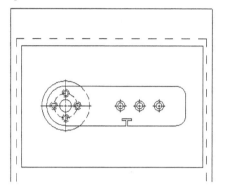

Creating Viewports in the ISO A1 Sheet

- Click the **ISO A1** tab below the graphics window.
- Select the viewport frame and modify the viewport using the grips, as shown below.

- Double-click inside the viewport to switch to the model space.
- Use the **Dynamic Zoom** and **Dynamic Pan** tools and drag the drawing to the center of the viewport.
- Click the **Annotation** drop-down and select the **2:1** from the menu.
- Use the **Dynamic Pan** tool and position the drawing, as shown in the figure.

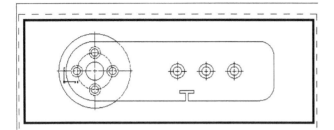

- Double-click outside the viewport to switch to the paper space.

- Use the **Circle** tool and create a 180 mm diameter circle on the Sheet, as shown below.

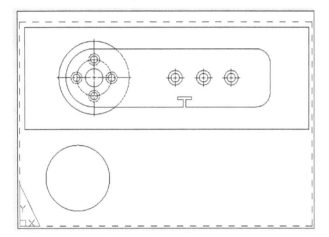

- Click **Sheet > Viewports > Viewport drop-down > Entity** on the ribbon.

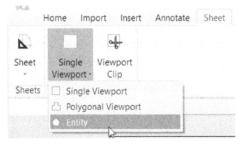

- Select the circle from the Sheet; it is converted into a viewport.

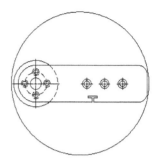

- Double-click in the circular viewport to switch to the model space.

- Click the **Annotation** drop-down on the status bar and select **4:1** from the menu; the drawing is zoomed in to its center.

- Use the **Dynamic Pan** tool and adjust the drawing, as shown below.

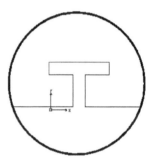

- Double-click outside the viewport.

- Click **Print > Print Preview** on the file menu; the plot preview is displayed. The viewport frames are also displayed in the preview.

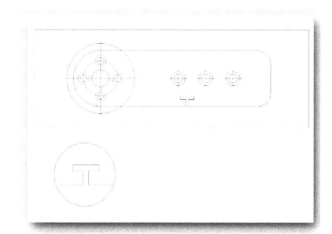

- Press ESC to close the preview window.

To hide viewport frames while plotting a drawing, follow the steps given below.

- Type **LA** in the command line to open the **Layers Manager**.
- In the **Layers Manager**, create a new layer called **Hide Viewports** and activate it.

- Deactivate the plotter symbol 🖶 under the **Print** column of the **Hide Viewports** layer; the object on this layer will not be printed.
- Click **OK** on the **Layers Manager**.
- Click the **Home** tab on the ribbon.
- On the **Layers** panel, click the **Active Layer** drop-down and select the **Entity to Active Layer** option.

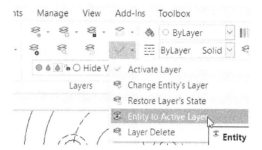

- Click the **Change to Current Layer** button on the **Layers** panel.

- Select the viewports in the **ISO A1** Sheet and press ENTER; the viewport frames become unplottable.
- To check this, click Print > Print **Preview** on the File menu; the plot preview is displayed as shown below.

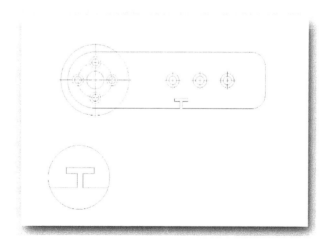

- Close the preview window.

Changing the Layer Properties in Viewports

The layer properties in viewports are not related to the layer properties in model space. You can change the layer properties in viewports without any effect in the model space.

- Double-click inside the larger viewport to activate the model space.
- Type **LA** in the command line to open the **Layers Manager**.
- In the **Layers Manager**, click the icon in the **Active ViewPort** column of the **Hidden** layer.
- Click **OK**; the hidden lines disappear in the viewport, as shown below.

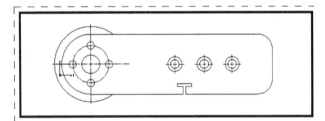

- Double-click outside the viewport to switch to paper space.
- Click the **Model** tab below the graphics window; the hidden lines are retained in the model space.

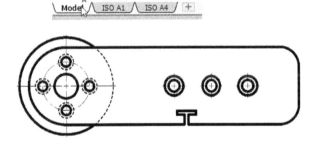

Creating the Title Block on the Sheet

You can draw objects on Sheets to create a title block, borders, and viewports. However, it is not recommended

to draw the actual drawing on Sheets. You can also create dimensions on Sheets.

Example1:

- Click the **ISO A1** Sheet tab.
- Activate the **Title Block** layer.
- Click **Polyline** drop-down > **Rectangle** on the **Draw** panel of the **Home** ribbon tab.
- Pick a point at the lower-left corner of the Sheet.
- Type **D** in the command line.
- Press ENTER.
- Type 820, and press ENTER.
- Type 550, and press ENTER; a rectangular border is created.
- Create a title block at the lower right corner, as shown below.

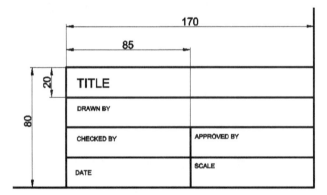

- Create attributes and place them inside the title, as shown below.

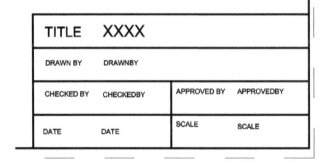

- Use the **Define Block** tool and convert it into a block.
- Use the **Insert Block** tool and insert it at the lower right corner of the Sheet.

- Save the drawing file as **Viewports-Example.dwg**.

Working with Annotative Dimensions

In DraftSight, you create drawings at their actual size. However, when you scale a drawing to fit inside a viewport, the size of the dimensions is not scaled properly. For example, in the following figure, the first viewport is scaled to 1:2, and the second viewport is scaled to 1:1. The dimensions in the first viewport appear much smaller.

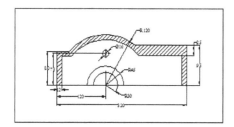

You can fix this problem by applying the Annotative property to dimensions.

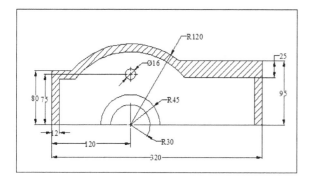

- Open the **Viewports-Example.dwg**, if not already opened.
- Set the **Dimensions** layer as current.
- Type **D** in the command line and press ENTER.
- In the **Drafting Styles** dialog, click the **New** button.
- In the **Create new DimensionStyle** dialog, enter **Dim_Anno** in the **Name** box and select **Annotative** from the **Based on** drop-down. Click **OK**.
- Expand **Lines > Dimension Line settings**.
- Type **1.25** in the **Offset** box.

- Expand the **Arrows** node and type 2.5 in the **Size** box.
- Expand **Text > Text Settings**.
- Type 2.5 in the **Height** box.
- Expand **Text > Text Position**.
- Select **Horizontal > Centered**.
- Select **Vertical > Centered**.
- Expand **Text > Text alignment**.
- Select the **Align Horizontally** option.
- Expand the **Linear Dimension** node.
- Select **Format > Decimal**.
- Select **Precision > 0**.
- Select **Decimal separator > '.'(Period)**.
- Click the **Activate** button.
- Click **OK** on the **Drafting Styles** dialog; the **Dim_Anno** style is listed in the **Dimension Style Control** drop-down on the **Dimensions** panel. Also, the annotation symbol is displayed next to it. This indicates that all dimensions created using this style will have annotative property.

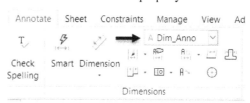

- Activate the **Linear Dimension** tool and set the Annotation Scale to **1:1**. Click **OK**.

- Create a linear dimension, as shown below.

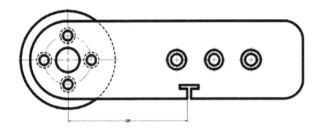

- On the status bar, click the **Annotation** drop-down and select **1:2**.

- On the ribbon, click **Annotate** > **Annotation Scaling** > **Annotation Scaling** drop-down > **Add Current Scale**.

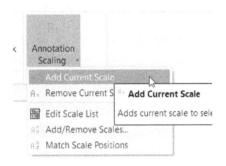

- Select the dimension from the graphics area.
- Press ENTER; the size of the dimension increases by two times.

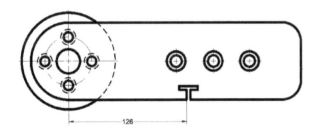

Example 2:

- Ensure that the **Annotation Scale** is set to **1:2** and create another linear dimension, as shown in the figure.

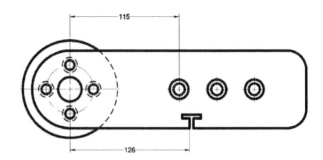

- Click the **ISO A4** Sheet in which the viewport scale is set to 1:2; the dimensions are scaled with respect to the viewport.

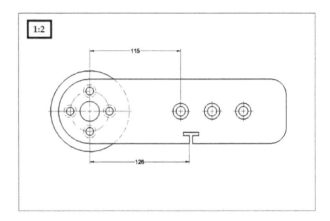

- Click the **ISO A1** Sheet; the dimensions are not displayed in the 2:1 viewport. To display dimensions in the 2:1 viewport, you need to add 2:1 scale to dimensions.

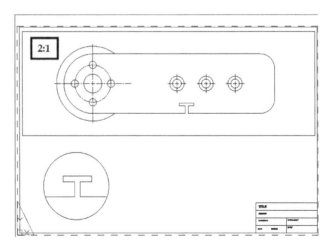

- Click the **Model** tab below the graphics window to switch to the model space.

- Click **Annotate > Annotation Scaling > Annotation Scaling drop-down > Add/Remove Scales** on the ribbon.

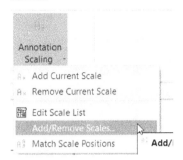

- Select the dimensions from the graphics window and right-click; the **Annotation Entity Scale List** dialog appears. In this dialog, the **Entity scale list** shows the scales applied to the selected dimensions. You need to add a 2:1 scale to the dimensions so that they are visible in the 2:1 viewport.

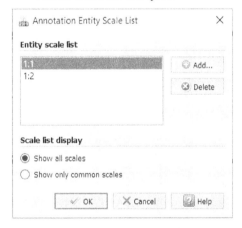

- To add a new scale to the dimensions, click the **Add** button; the **Scale List** dialog appears.
- Select the **2:1** scale from the list and click **OK**; the scale is added to the **Entity scale list**.

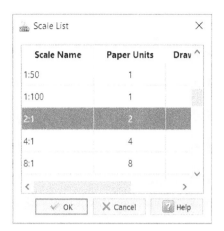

- Click **OK** on the **Annotation Entity Scale List** dialog.
- Click the **ISO A1** Sheet; the dimensions are displayed in both 2:1 and 1:2 viewports.

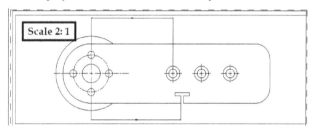

- Similarly, create other dimensions, as shown below. Add 2:1 and 1:2 scales to dimensions and check the drawing in two different Sheets.

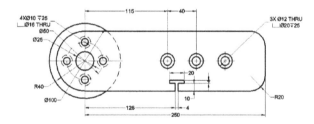

Scaling Hatches relative to Viewports

While working in Sheets, you may also need to scale the hatch with respect to the viewport scale. The following figure shows a drawing in two different viewports 1:2 and 1:1. The hatch in the left viewport is smaller than that in the right side viewport. You can correct this problem by using the **Annotative Scaling** option.

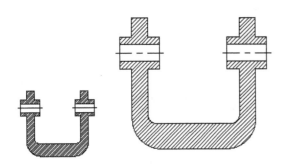

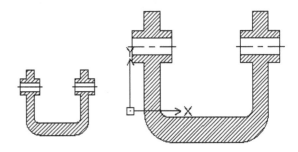

- Click the Model tab below the graphics area.

- Select the hatch patterns from the drawing.

- Right click and select **Hatch Edit**; the **Hatch/Fill** dialog appears.

- In the **Hatch/Fill** dialog, check the **Annotative scaling** option in the **Mode** section.

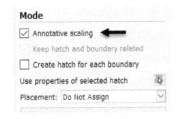

- Click **OK**.

- Select the hatch pattern.

- Right-click and select **Annotative Entity Scale > Add/Remove Scales**.

- Click the **Add** button on the **Annotation Entity Scale list** dialog.

- Select the 1:2 scale from the **Scale list** dialog.

- Click **OK** twice.

- View the two different sheets and notice that the hatch is scaled with respect to the viewport scale.

Working with Annotative Text

Annotative property can also be assigned to text. The annotative text is scaled with respect to the viewport scale.

- Open the **Viewports-Example.dwg**, if not already opened.

- Click **Annotate > Text > Text Style** on the ribbon; the **Drafting Styles** dialog appears.

- Click the **New** button on the **Text Style** dialog; the **Create new TextStyle** dialog appears.

- Enter **Text_Anno** and click **OK**.

- Select the **Text_Anno** style from the **Style** list.

- Set **Font** to **Arial**.

- Select the **Annotative Scaling** checkbox in the Height section.

- Set **Sheet Text Height** to **2.5**.

- In the **Orientation** section, set the **Spacing** to **1**.

- Click **Apply** and **OK**.

- Select **1:1** from the **Annotation** menu on the status bar.

- Click **Annotate > Text > Note** on the ribbon.

- Specify the first corner of the text editor by picking an arbitrary point.

- Type J in the command line, and then press ENTER; the command line displays:

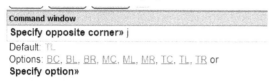

- Type **MC** in the command line, and then press ENTER.

- Move the pointer toward the right and specify the second corner of the text editor.

- Type **All dimensions are in mm** and click the **Create Note** button on the **Note Formatting** panel.

- Move the text and place it at the bottom left corner of the drawing, as shown below.

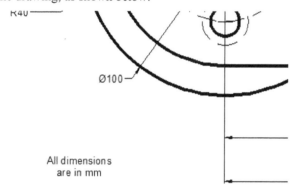

- View the drawing in the **ISO A4** Sheet; the text is not displayed. This is because the text is set to a 1:1 scale.

- On the status bar, click the **Annotation** drop-down and deselect the **Show annotation entities for current scale only** option.

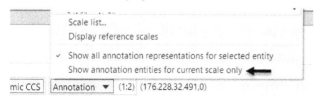

The text is visible in the ISO A4 Sheet.

- Save the drawing as **Sheet Example.dwg** and close.

Creating Templates

After specifying the required settings in a drawing file, you can save those settings for future use. You can do so by creating a template. Template files have settings such as units, limits, and layers already created, which will increase your productivity. In previous sections, you have configured various settings, such as layers, colors, linetypes, and plotting settings. Now, you create a template file containing all of these settings and the title block that you have created.

- Click the **New** button on the **Quick Access Toolbar**; the **Select Template** dialog appears.

- Select **Open > Open with no Template – Metric** from the bottom right corner of the dialog; a drawing file is opened.

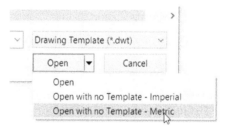

- Open the **Layers Manager** and create the layers contained in the table below:

Layer	Linetype	Lineweight
Construction	Continuous	Default
Object	Continuous	0.7 mm
Hidden Lines	Hidden	0.3 mm
Center Lines	CENTER	0.25 mm
Dimensions	Continuous	0.25 mm
Section Lines	Continuous	0.5 mm
Cutting Plane	Phantom	0.6mm

Title Block	Continuous	1mm
Viewport	Continuous	0.25 mm
Text	Continuous	Default
Title block text	Continuous	Default

- Click the **Sheet 1** tab to activate the paper space.
- Click the right mouse button on the **Sheet 1** tab and select **Print Configuration Manager**; the **Print Configuration Manager** dialog appears.
- Click **Modify** on the **Page Setup** Manager; the **Page Setup** dialog appears.
- In the **Print Configuration Manager** dialog, click the **New** button; the **New Print Configuration** dialog appears.

- Select the **Default** option and click **OK**.
- Type **ISO A3** in the **File name** box
- Click **Save**.
- Under the **Printer/plotter** group, select the plotter that you have configured to your workstation.
- Set the **Paper Size** to **ISO A3 (420 x 297 MM)**.
- Click **Save.**
- Select the **ISO A3** option from the list.
- Click the **Activate** button.
- Click **Close** to exit both the dialogs.
- Delete the Viewport from the sheet.
- On the ribbon, click **View** tab > **Navigate** panel > **Zoom drop-down** > **Zoom Bounds**.
- Draw a title block in the paper space, as shown below.

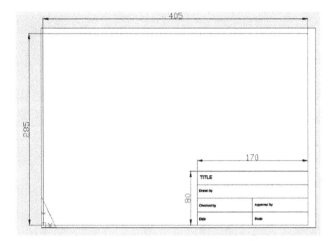

- Create a viewport inside the title block (refer to the **Creating Viewports in the Paper space** section discussed earlier in this chapter).

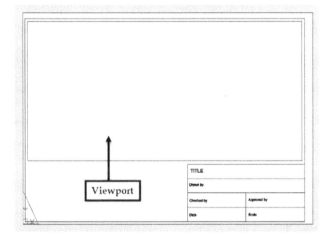

- On the **Quick Access toolbar**, click the **Save** button; the **Save As** dialog appears.
- In the **Save As** dialog, set **Save as type** to **Drawing Template (*.dwt).**
- Enter **ISOA3** in the **File name** box and click **Save**.

Plotting/Printing the drawing

- Click the **New** button on the Quick Access Toolbar.
- Select **ISOA3** from the **Specify Template** dialog.
- Click Open
- A new drawing is started with the selected template.
- Open the **Layers Manager**; notice that the layers saved in the template file are loaded automatically.

- Close the **Layers Manager**.
- Create a drawing, as shown below. You can also download the drawing from the companion website.

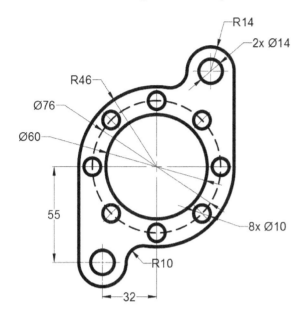

- Click the **Sheet 1** tab to activate the paper space.
- Double-click inside the viewport to activate the model space.
- Set the **Annotation scale** to 1:1 on the status bar.
- Use the **Pan** tool and position the drawing in the center of the viewport.
- Double-click outside the viewport to activate the paper space.
- Hide the viewport frame by freezing the **Viewport** layer.

- Click the **Print** button on the **Quick Access Toolbar**;

the **Print** dialog appears.

- Make sure that the options in this dialog are the same as that you specified while creating the template.
- Click the **Print Preview** button located at the bottom left corner; the preview window appears.
- Click the **Zoom Original** button to fit the drawing to the window.

- Examine the print preview for the desired output and click the **Plot** button; the drawing is printed.

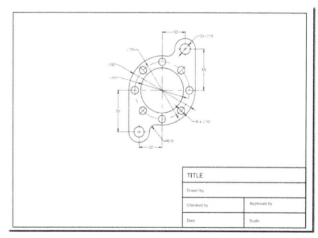

- Save and close the drawing file.

Exercise 1

Create the drawing, as shown below. After creating the drawing, perform the following tasks:

- Create a Sheet of the A3 size and then create a viewport.

- Set the viewport scale to 1:2.

- Set the scale of the dimensions and hatch lines with respect to the viewport.

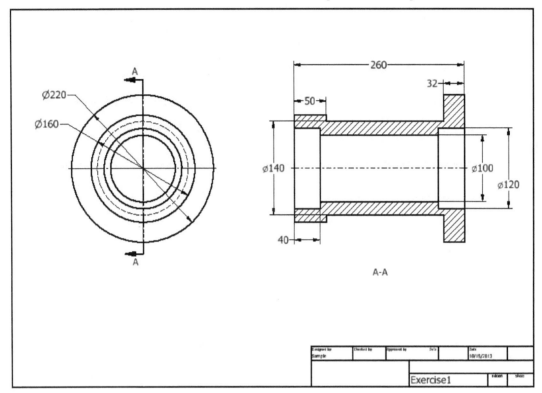